SpringerBriefs in Environmental Science

SpringerBriefs in Environmental Science present concise summaries of cutting-edge research and practical applications across a wide spectrum of environmental fields, with fast turnaround time to publication. Featuring compact volumes of 50 to 125 pages, the series covers a range of content from professional to academic. Monographs of new material are considered for the SpringerBriefs in Environmental Science series.

Typical topics might include: a timely report of state-of-the-art analytical techniques, a bridge between new research results, as published in journal articles and a contextual literature review, a snapshot of a hot or emerging topic, an in-depth case study or technical example, a presentation of core concepts that students must understand in order to make independent contributions, best practices or protocols to be followed, a series of short case studies/debates highlighting a specific angle.

SpringerBriefs in Environmental Science allow authors to present their ideas and readers to absorb them with minimal time investment. Both solicited and unsolicited manuscripts are considered for publication.

More information about this series at http://www.springer.com/series/8868

Eldiiar Duulatov • Xi Chen • Gulnura Issanova
Rustam Orozbaev • Yerbolat Mukanov
Amobichukwu C. Amanambu

Current and Future Trends of Rainfall Erosivity and Soil Erosion in Central Asia

 Springer

Eldiiar Duulatov
Institute of Geology
National Academy of Sciences of the
Kyrgyz Republic
Bishkek, Kyrgyzstan

Xi Chen
State Key Laboratory of Desert & Oasis
Ecology, Xinjiang Institute of Ecology and
Geography
Chinese Academy of Sciences
Urumqi, Xinjiang, China

Gulnura Issanova
Al-Farabi Kazakh National University
Almaty, Kazakhstan

Rustam Orozbaev
Institute of Geology
National Academy of Sciences of the
Kyrgyz Republic
Bishkek, Kyrgyzstan

Yerbolat Mukanov
Department of Agro-meteorological
Monitoring and Forecasting
RSE Kazhydromet
Nur-Sultan, Kazakhstan

Amobichukwu C. Amanambu
Spatial and Temporal Analysis of Rivers
(STAR) Lab, Department of Geography
University of Florida
Gainesville, FL, USA

ISSN 2191-5547   ISSN 2191-5555   (electronic)
SpringerBriefs in Environmental Science
ISBN 978-3-030-63508-4   ISBN 978-3-030-63509-1   (eBook)
https://doi.org/10.1007/978-3-030-63509-1

This Springer imprint is published by the registered company Springer Nature Switzerland AG
The registered company address is: Gewerbestrasse 11, 6330 Cham, Switzerland

Preface

The resilience of soil resources in Central Asia is one of the main problems in the context of climate change. Central Asia is included in the list of regions that are very vulnerable to the adverse effects of climate change. Climatic factors largely control rainfall erosivity and soil erosion. Consequently, changes in precipitation can affect the spatial distribution of soil. Climate change is expected to affect erosivity in Central Asia and other regions of the world.

The relationship between precipitation characteristics and erosion has been established. It is essential to know the temporal and spatial differences of erosion from rainfall erosivity. Therefore, it makes sense to interpret whether the tendency to reduce precipitation, for the historical and future climate in Central Asia, will have consequences for the erosive power of precipitation. The spatial and temporal projection of future rainfall erosivity in a changing climate in Central Asia has not been studied. Assessing the rainfall erosivity and its consequences can assist specialists and researchers express best practices for soil conservation. The result of this type of research is all-encompassing and may reflect the normal variations in other parts of the world (e.g., the arid and semi-arid regions) and not inherently be limited to the field of research.

The main purpose of this book is to analyse the climate change influence on rainfall erosivity and soil erosion across Central Asia. This will be achieved primarily using the RUSLE model with past and future climate projections, spatiotemporal variations of rainfall erosivity and soil erosion based on WorldClim and Coupled Model Intercomparison Project Phase 5 (CMIP5) climate models (for Central Asia and separately Kazakhstan).

The book is mainly addressed to scientists and researchers whose studies have focused on studying the climate change, rainfall erosivity and soil erosion, as well as students and planners. We believe that this publication makes a great contribution to our knowledge about climate change, soil erosion and rainfall erosivity over Central Asia.

Bishkek, Kyrgyzstan   Eldiiar Duulatov
Urumqi, Xinjiang, China   Xi Chen
Almaty, Kazakhstan   Gulnura Issanova
Bishkek, Kyrgyzstan   Rustam Orozbaev
Nur-Sultan, Kazakhstan   Yerbolat Mukanov
Gainesville, FL, USA   Amobichukwu C. Amanambu

Acknowledgements

Many scientists have contributed, supported and provided the basis for this book.

The authors sincerely thank Friday U. Ochege, Gulkaiyr Omurakunova, Merim Pamirbek kyzy and Talant Asankulov.

We are also very grateful to all our colleagues at Institute of Geology, National Academy of Sciences of the Kyrgyz Republic, in particular Kadyrbek Sakiev, Salamat Alamanov, Aizek Bakirov and Salamat Abdyzhapar uulu. At Xinjiang Institute of Ecology and Geography, Chinese Academy of Sciences, in particular Jilili Abuduwaili, Bao Anming, Chen Yaning, Liu Tie and Li Yaoming.

Thanks to all our colleagues at the Research Center for Ecology and Environment of Central Asia (Bishkek).

The authors thank the aforementioned contributors, colleagues, friends and well-wishers for their support and engagement.

The authors would like to express their sincere gratitude to the WorldClim data portal, Climate Change Agriculture and Food Security (CCAFS), National Aeronautics and Space Administration (NASA) and the National Snow and Ice Data Center (NSIDC) for providing climate and remote sensing data.

This research was funded and supported by the Chinese Academy of Sciences President's International Fellowship Initiative (2020PC0090) and the Foundation State Key Laboratory of Desert and Oasis Ecology, Xinjiang Institute of Ecology and Geography, Chinese Academy of Sciences; the Pan – Third Pole Environment Study for a Green Silk Road (Grant No. XDA20060303); the International Cooperation Project of the National Natural Science Foundation of China (Grant No. 41761144079); and the Strategic Priority Research Program of the Chinese Academy of Sciences (Grant No. XDA20060301).

Content and Structure of the Book

This book summarizes and provides the outcomes of research results and recent studies related to soil erosion in Central Asia, where current and future trends in rainfall erosivity and erosivity density are the greatest, and discusses the potential impact on the environment across the region.

The book has six individual chapters as follows. The first chapter, "Introduction and Background on Rainfall Erosivity Processes and Soil Erosion", provides an overview of background on rainfall erosivity processes and soil erosion. It considers, evaluates and analyses global climate scenario data, the predicted impact of climate change on rainfall erosion along with the rainfall erosion models.

The second chapter, "Natural Conditions of Central Asia", provides information about the topography, climate conditions and description of soils in Central Asia.

The third chapter, "Data Sources and Methodology", analyses a detailed methodology on the RUSLE model, predicting the impact of climate change on precipitation erosion based on geographic information system GIS and remote sensing (RS) techniques to assess soil erosion.

The fourth chapter, "Projected Rainfall Erosivity and Soil Erosion in Central Asia", provides analyses on projected changes on rainfall erosivity over Central Asia under RCPs 2.6 and 8.5 for two periods (2030s and 2070s).

The fifth chapter, "Spatio-temporal Variations and Projected Rainfall Erosivity and Erosivity Density in Kazakhstan", analyses an annual trend of precipitation erosion in Kazakhstan for the period 1970–2017 and estimates long-term changes in annual precipitation erosion in Kazakhstan using past and future climate data and three GCM scenarios and two RCPs for three future periods.

The sixth chapter "Conclusions and Recommendations", this chapter concludes the preceding chapters and offers recommendations for carrying out future studies.

Contents

About the Authors

Eldiiar Duulatov is a research associate in the Geography Department at Institute of Geology, National Academy of Sciences of the Kyrgyz Republic (NAS KR), and senior lecturer in the Geography, Ecology and Tourism Department, Jusup Balasagyn Kyrgyz National University. He received his Ph.D. in Cartography and GIS in 2019. Dr. Duulatov studied at the Kyrgyz National University for a bachelor's degree (B.Sc.) in human geography and a master's (M.Sc.) degree in physical geography and at Xinjiang Institute of Ecology and Geography, University of Chinese Academy of Sciences, China, for his doctoral degree. His research interests were focused on problems of soil erosion, in particular, the climate change–induced rainfall erosivity in the soil erosion processes during a Ph.D. study. Currently, he is working on water resources and hydrological modelling in the upper stream of the Syrdarya Basin. Dr. Duulatov participates regularly in the International Scientific Activities (Conference, Forum, and Symposium) on Environmental Problems as well as writes articles on the subject and takes part in local and international projects. He has published several SCI papers in high-ranking international peer-reviewed journals and is an active member of the Kyrgyz Geographical Society.

Xi Chen is a professor of hydrology, ecosystem, remote sensing and geographic information system of arid land; general director of the Research Center for Ecology and Environment of Central Asia, Chinese Academy of Sciences; and vice president of the Xinjiang branch of the Chinese Academy of Sciences. He has made a great contribution by developing geography and ecosystem management theories and methods in arid regions and published more than 200 papers and 18 books. In 2008, Prof. Chen established the ecosystem monitoring and research network of Central Asia; in 2009, he founded the *Journal of Arid Land* and is the editor in chief since then. In 2012, he was elected as the chair of the Central Asia Ecology and Environment Research Alliance, carrying out more than 30 international cooperation projects on climate change, ecosystem, and agriculture and water resource with Kazakhstan, Kyrgyzstan, Tajikistan, Uzbekistan, Germany and Russia. Prof. Chen is an honorary academician at the National Academy of Sciences of the Kyrgyz Republic and the Republic of Kazakhstan.

Gulnura Issanova holds a doctoral degree in natural sciences and is an associate professor at Al-Farabi Kazakh National University, scientist and researcher at U.U. Uspanov Kazakh Research Institute of Soil Science and Agrochemistry, and a scientific secretary at the Research Centre of Ecology and Environment of Central Asia, Almaty, Kazakhstan. She studied at Al-Farabi Kazakh National University for a bachelor's degree (B.Sc.) and master's (M.Sc.) degree in physical geography and at Xinjiang Institute of Ecology and Geography, Chinese Academy of Sciences, China, for her doctoral degree. She completed a postdoc under the CAS President's (Bai Chunli) International Fellowship Initiative (PIFI) for 2017–2018 at the Xinjiang Institute of Ecology and Geography, Chinese Academy of Sciences, China. Dr. Issanova is a holder of the Foreign Expert Certificate of the People's Republic of China. Currently, she is a postdoc at Al-Farabi Kazakh National University, Faculty of Geography and Environment. Her research interest is focused on problems of soil degradation and desertification, in particular, the role of dust and sand storms in the processes of land and soil degradation and desertification during her Ph.D. study. Currently, she is interested in and focusing on water resources and lakes (availability, state and consumption/ use) and Aeolian processes in Central Asian countries. Dr. Issanova has published many SCI papers in international peer-reviewed journals with a high level and wrote a handbook, *How to Write Scientific Papers for International Peer-Reviewed Journals* (in Russian and Kazakh languages). She is an author and co-author of several monographs: *Aeolian processes as dust storms in the deserts of Central Asia and Kazakhstan* (2017), *Man-Made Ecology of East Kazakhstan* (2018), *Hydrology and Limnology of Central Asia* (2019), *Geomechanical Processes and Their Assessment in the Rock Massifs in Central Kazakhstan* (2020) by Springer Nature. She became a laureate of the international award "Springer Top Author" and was awarded in the Nomination "Springer Young Scientist Awards-2016" for high publication activity in scientific journals published by Springer Nature. Dr. Issanova was included in the list of "Top-8 Young Scientists and Top-18 Leading Scientists from Kazakhstan" in 2016.

Rustam Orozbaev is a senior research scientist and an acting director at the Institute of Geology, National Academy of Sciences of the Kyrgyz Republic. He obtained his master's (M.Sc., 2005) and doctoral (D. Sc., 2010) degrees in geology from the Shimane University (Japan). Dr. Orozbaev studied geology, geochemistry and the nature and dynamics of deep fluids in subduction zones at Kyoto University (Japan) as Japan Society for the Promotion of Science (JSPS) postdoctoral research fellow from 2011 to 2014. His research areas are geology, petrology, tectonics, mining and environmental geology of the Tien-Shan Mts. Dr. Orozbaev received several awards from the Kyrgyz National Academy of Sciences and the Geochemical Society of Japan for his scientific achievements and published papers.

Yerbolat Mukanov holds a doctoral degree in natural sciences and is a scientist and researcher at Xinjiang Institute of Ecology and Geography, CAS, and the head of the Agro-meteorological Forecasting Division in the Department of Agrometeorological Monitoring and Forecasting at RSE Kazhydromet (Nur-Sultan), Kazakhstan. He obtained his bachelor's degree from the Agrarian University of S. Seifullin, Astana, Kazakhstan; master's degree from Nanjing University Information Science & Technology, Nanjing, China; and doctoral degree at Xinjiang Institute of Ecology and Geography, Chinese Academy of China, Urumqi, China. His research interests are focused on agriculture, climate change, hydrology, soil degradation and desertification.

Amobichukwu C. Amanambu is a PhD candidate at the Department of Geography, University of Florida, USA. Obtained his bachelor's and master's degrees from the prestigious University of Ibadan, Nigeria, where he studied geography. Amobichukwu is currently studying for another master's degree in water and environmental management at Loughborough University in the United Kingdom. His research focus is on watershed hydrology, which includes flooding, erosion, water resources and river processes. He enjoys applying methodologies from remote sensing and GIS in many of his research works. Amobichukwu is an author and co-author of many research articles in high-profile journals. He is an active member of many international organizations including the American Association of Geographers (AAG), International Association of Hydrological Society (IAHS) and American Society for Photogrammetry and Remote Sensing (ASPRS).

Abbreviations and Acronyms

AR5	Fifth Assessment Report
ASL	Above Sea Level
°C	Degree Celsius
CCAFS	Climate Change Agriculture and Food Security
CMIP5	Coupled Model Inter-comparison Project Phase 5
DEM	Digital Elevation Model
DSMW	Digital Soil Map of the World
ENSO	El Niño Southern Oscillation
ESACCI-LC	European Space Agency Climate Change Initiative Land Cover
FAO	Food and Agriculture Organization
GCM	Global Climate Model
GDP	Gross Domestic Product
GIS	Geographic Information Systems
HWSD	Harmonized World Soil Data
IPCC	Intergovernmental Panel on Climate Change
Kazhydromet	Hydro-Meteorological Agency of Kazakhstan
LULC	Land Use and Land Cover
MC	Mean Correlation
MS	Mean Sensitivity
MSE	Mean Square Error
MODIS	Moderate Resolution Imaging Spectroradiometer
MUSLE	Modified Universal Soil Loss Equation
NAO	North Atlantic Oscillation
NASA	National Aeronautics and Space Agency
NCDC	National Climatic Data Center
NDVI	Normalized Difference Vegetation Index
NOAA	National Oceanic and Atmospheric Administration
NSE	Nash-Sutcliffe Efficiency Coefficient
NSIDC	National Snow and Ice Data Center
R	Correlation Coefficient
R^2	Coefficient of Determination

RCP	Representative Concentration Pathway
RMSE	Root Mean Square Error
RS	Remote Sensing
RUSLE	Revised Universal Soil Loss Equation
SAGA	System for Automated Geoscientific Analyses
SD	Standard Deviation
SST	Sea Surface Temperature
SWAT	Soil and Water Assessment Tool
SRTM	Shuttle Radar Topography Mission
USDA	United States Department of Agriculture
USGS	United States Geological Survey
USLE	Universal Soil Loss Equation
WorldClim	World Climate Data

Chapter 1
Introduction and Background of Rainfall Erosivity Processes and Soil Erosion

Abstract Climate change-induced precipitation variability is the leading cause of rainfall erosivity that leads to excessive soil losses in most countries of the world. Central Asia is included in the list of regions that are very vulnerable to the adverse effects of climate change. Climatic factors largely control rainfall erosivity and soil erosion. Consequently, changes in precipitation can affect the spatial distribution of the soil. Climate change is expected to affect erosivity in Central Asia and other regions of the globe. This chapter provides an overview of background on rainfall erosivity processes and soil erosion. It considers, evaluates, and analyses global climate scenario data, the predicted impact of climate change on rainfall erosivity along with the soil erosion models.

Keywords Rainfall erosivity · Soil erosion · Central Asia · Climate change · Tien-Shan · Precipitation · Impact · CMIP5 · GCMs · RUSLE · GIS · RS · Spatiotemporal variability · Spatiotemporal characteristics

1.1 Rainfall Erosivity Processes

Rainfall erosivity, the R factor, which combines the influence of the duration, magnitude, and intensity of precipitation, is vital for many soil erosion models (Lai et al. 2016). It cannot be modified by humans, and it differs from soil characteristics, vegetation cover, and soil conservation (Angulo-Martínez and Beguería 2009). Rainfall erosivity is concerned with the potential ability of precipitation to cause erosion, so it reflects the risk of soil erosion when the condition of the underlying surface is unchanged for the region (Gu et al. 2018). In the Revised Universal Soil Loss Equation (RUSLE) model, rainfall erosivity is initially defined as the precipitation energy multiplied by the maximum 30-min precipitation intensity (EI_{30}), and the annual R is the sum of EI_{30}, which is calculated from the recorded precipitation (Renard et al. 1997; Wischmeier and Smith 1978).

E. Duulatov et al., *Current and Future Trends of Rainfall Erosivity and Soil Erosion in Central Asia*, SpringerBriefs in Environmental Science, https://doi.org/10.1007/978-3-030-63509-1_1

However, the main disadvantage of the RUSLE R factor is that it needs continuous data on precipitation. Data on the pluviograph for at least 20 years is required to calculate the initial rainfall erosivity (Renard et al. 1997). Data with such a high temporal resolution is not available in many countries and regions, and its processing is very tedious and time-consuming (Lai et al. 2016; Lee and Heo 2011). Therefore, numerous studies have established a mathematical regression equation between R and variable rainfall, such as annual rainfall (Renard and Freimund 1994), monthly rainfall, and daily rainfall (Panagos et al. 2016b). These simplified methods provide great convenience for studying the spatial and temporal variabilities of rainfall erosivity. Annual precipitation data are relatively easy to obtain in most places and are reliable, and this simplified method assumes that annual erosivity is correlated with annual precipitation (Lee and Heo 2011). Annual precipitation data were used as simple estimates of rainfall erosivity in many parts of the globe (Almagro et al. 2017; Amanambu et al. 2019; Duulatov et al. 2019; Lee and Heo 2011; Naipal et al. 2015; Renard and Freimund 1994; Yang et al. 2003).

Rainfall erosivity is the ability of rainfall to cause soil erosion through raindrop impact and surface washout when the infiltration capacity is exceeded (Almagro et al. 2017). It has attracted considerable attention in the process of quantitative prediction of soil erosion and sediment yield. Among all the factors of erosion, rainfall erosivity and land cover/control factor are considered to be the most dynamic (Panagos et al. 2016a).

1.2 Impacts of Climate Change on Rainfall Erosivity

Climate changes that are related to soil erosion mainly include changes in temperature and precipitation (Li and Fang 2016). Climate change may change rainfall erosivity due to the changes in precipitation patterns (Mondal et al. 2016). The characteristics of precipitation (amount of precipitation, intensity, and spatial-temporal distribution) directly cause soil erosion (Teng et al. 2018). Conversely, an increase in temperature indirectly causes soil erosion (Li and Fang 2016). The addition of water vapor to the atmosphere affects the nature of climate circulation, thereby changing the intensity and frequency of extreme precipitation (Almagro et al. 2017).

In arid and semi-arid climates, such as in Central Asia, temperature and rainfall events increases more significantly than those in many other regions worldwide (Immerzeel et al. 2013; Unger-Shayesteh et al. 2013). According to the Intergovernmental Panel on Climate Change (IPCC) Fifth Assessment Report (AR5), the global average precipitation and surface temperature have significantly changed, and the report assumes that these changes are likely to continue throughout the twenty-first century (Change 2014). Numerous researchers (Almagro et al. 2017; Amanambu et al. 2019; Borrelli et al. 2020; Gupta and Kumar 2017; Plangoen et al. 2013; Teng et al. 2018; Yang et al. 2003) have described the impact of climate change on soil erosion by water globally and regionally.

Future potential changes in rainfall erosivity can be projected using the precipitation data obtained from general climate models (GCMs) under different greenhouse gas emission scenarios that follow the various Representative Concentration Pathways (RCPs) of CMIP5. More than 50 GCMs are currently available for environmental studies (Panagos et al. 2017a). The use of GCMs has indicated that the changes in precipitation patterns impact soil erosion in several ways, as well as the changes in rainfall erosivity (Plangoen et al. 2013). Several scholars have separately studied the changes in erosivity (Almagro et al. 2017) and also conducted combined studies on erosivity and soil erosion (Amanambu et al. 2019; Teng et al. 2018) in different parts of the globe using GCMs. The changes in rainfall erosivity were directly related to the impact of climate change on soil erosion (Nearing 2001), as the levels of erosivity are expected to change in response to climate change (Amanambu et al. 2019).

1.3   Soil Erosion Processes and Models

Soil is a fragile resource that requires time to recover. Without soil, agricultural production is inconceivable, and the improvement of the well-being of the people is impossible (Mamytov and Roychenko 1961). Soils are subject to water erosion, which leads to their degradation (Riquetti et al. 2020). Today, soil erosion is one of the most severe environmental problems that the world is facing. Soil erosion by water is by far the most important type of soil erosion, which affects ~125 million km^2 of land surface worldwide (Borrelli et al. 2017). Among the continents, Asia ranks third in terms of the severity of soil erosion (Borrelli et al. 2017). However, with the changing climate, the problem of soil erosion could be exacerbated and must be addressed.

Soil erosion is the combination of natural and anthropogenic processes that cause changes in soil functions in the geosystem; quantitative and qualitative degradation of soil composition, properties, and regimes; and decline in the natural and economic significance of lands (Khitrov et al. 2007). Soil degradation also causes huge economic damage, disrupts the ecological balance, and worsens the social conditions of the people's lives (Dobrovolsky 2002). The scientifically grounded and rational use of lands largely depends on the correct identification and establishment of the degree or category of erosion of the soil cover and accurate accounting for their correct nomenclature and classification.

Land degradation is a global environmental crisis that is threatening agricultural areas at an alarming rate. Global communities are mostly aware of this crisis, along with energy crisis and global warming, as it directly impacts the production of food for people. Land degradation occurs when natural or human-made processes reduce the ability of the earth to support crops, livestock, and organisms. The soil is the earth's fragile skin, without which life on earth will be unsustainable (Pimentel 2006).

Agriculture is the leading economic resource in Central Asia. It accounts for 10–45% of its gross domestic product (GDP), which employs between 20 and 50%

of the workforce (Qushimov et al. 2007). Currently, agriculture remains an important sector in the economy of Central Asia, providing 5.2% of the GDP in Kazakhstan, 7.5% in Turkmenistan, 18.5% in Uzbekistan, 20.8% in Kyrgyzstan, and 23.3% in Tajikistan (Hamidov et al. 2016). Consequently, a better understanding of the impact of climate change on soil erosion is also essential to improve the economy of Central Asia.

Along with the atmosphere and hydrosphere, the soil cover (pedosphere) is also subjected to negative anthropogenic influences (Denisov 2006; Shigaeva 2008). The protection of soils and their rational use are of primary importance for the financial and social growth of the country. The significance of the current state of soil resources, their judicious use, and their careful treatment help improve their fertility. Soil erosion is adapted to the biophysical environment, including soil, sediments, topography, land cover, and interactions between them.

The major features of terrains affecting the mechanism of soil loss are slope length, shape, and appearance. The impact of slope and aspect plays a significant role in the flow mechanism. The higher the slope, the higher the runoff, and the more infiltration is decreased. The drain formed on the slope finds a path nearby, which causes soil erosion as the flow rate increases.

Due to erosion occurring over the past 40 years, 30% of the arable land worldwide has become unproductive (Jahun et al. 2015). Erosion occurs when the soil remains exposed to rain or wind energy, so raindrops with high energy hit the open soil and easily remove soil particles from the surface. The impact intensifies on sloping areas, where often more than half of the soil surface is carried away as water flows along valleys and waterways (Jahun et al. 2015). Thus, the rate of erosion is influenced by soil composition, land slope, and vegetation cover. The problem of population growth is the increased demand for food and arable land. As a result, forest, soil, and water resources are exploited wastefully. Soil erosion also causes off-site damage, such as river sediment deposition, reservoir sedimentation, and canal siltation. Damages can lead to significant economic losses (Li and Fang 2016).

Soil structure influences the ease with which it can be eroded. Soils with medium to fine texture, low organic matter content, and weak structural development are easily eroded. The erodibility of soil depends on its texture and structure, organic matter content, and permeability (Wischmeier and Smith 1978). Soil erodibility refers to the susceptibility of the soil to erosion. At a general level, this depends primarily on the structural stability of the soil and its ability to absorb rainfall.

The topography of a given landscape, its rainfall, and wind all combine to influence its susceptibility to erosion. Topography is undoubtedly one of the most critical determinants of soil erosion. Typically, erosion only becomes severe when the slope angle exceeds a critical steepness and then subsequently increases logarithmically. Runoff and erosion also tend to increase with the increase in slope length. The effect of topography also often occurs at a mostly local level, with erosion being initiated in specific locations on the slope, or in association with minor topographic variations.

In practical terms, the vegetation cover is possibly the most crucial element in the hillslope erosion model, since it is the factor that can be readily altered. Moreover, it provides the first opportunity for soil erosion control (Briggs et al. 1992). Land areas covered by plant biomass are more protected and experience relatively little soil erosion because the biomass layer dissolves the raindrop, and the wind energy and topsoil are held by the biomass (Pimentel 2006).

Even though the natural resources of the modern world are still not fully mapped, the level of mapping, growth potential, and demand make the understanding of the relationships between various natural components and their quantification to predict possible scenarios, that is, modeling, more and more exciting and vital. Nevertheless, the human impact on the ecosystem is another unpredictable component that makes modeling so accessible and thus should be taken into account.

Soil erosion and sedimentation by water includes the processes of detachment, transportation, and deposition of sediments under the influence of a raindrop and running water (Foster and Meyer 1977; Wischmeier and Smith 1978). The main forces come from the effects of raindrops and running water (Fig. 1.1). Soil erosion by water depends not only on anthropogenic factors but also on physiographic factors (e.g., rainfall intensity, runoff, topography, and soil texture). Unsurprisingly, due to the significance of this problem (soil erosion by water), numerous studies about it have been conducted worldwide. It is important to quantify the impacts of soil erosion by water and to develop effective measures for soil and water conservation.

Numerous models have been developed to predict the process of degradation. However, these are mostly water erosion prediction models and are well known, e.g., Water Erosion Prediction Project (WEPP) (Flanagan and Nearing 1995), Soil and Water Assessment Tool (SWAT) (Arnold et al. 1998), Universal Soil Loss Equation (USLE) (Wischmeier and Smith 1978), PAN-European Soil Erosion Risk Assessment (PESERA) (Kirkby et al. 2004), and European Soil Erosion Model (EUROSEM) (Morgan et al. 1998). For other types of degradation, there is also a large number of different models, such as NUTMON (Smaling and Fresco 1993) for the balance of nutrients, Agricultural Production Systems Simulator (APSIM)

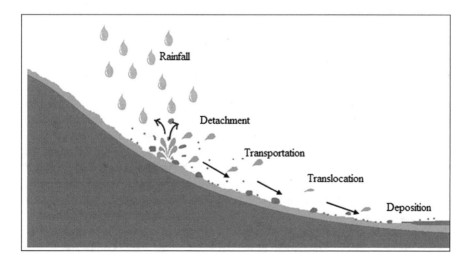

Fig. 1.1 Mechanism of soil erosion (Zafirah et al. 2017)

(McCown et al. 1996) for predicting productivity, WOFOST as a simulation model of crop production (Van Diepen et al. 1989), and SOVEUR (Batjes 1997) for predicting pollution.

The USLE model was first proposed by Wischmeier and Smith (1965) based on the concept of separation and transport of particles from precipitates to calculate the degree of soil erosion in agricultural areas. The equation was improved in 1978. This is the most widely used and generally accepted empirical model of soil erosion, which was developed for sheet and rill erosion. USLE has been improved over the past 40 years by many researchers (Alewell et al. 2019).

Modified Universal Soil Loss Equation (MUSLE) (Williams 1975), Revised Universal Soil Loss Equation (RUSLE) (Renard et al. 1997), Areal Nonpoint Source Watershed Environmental Resources Simulation (ANSWERS) (Beasley et al. 1980), and Unit Stream Power-Based Erosion Deposition (USPED) (Mitasova et al. 1996) are based on the USLE and is an improvement of the former. Among these models, the RUSLE was widely used to assess the long-term rate of soil erosion in large-scale studies (Amanambu et al. 2019; Naipal et al. 2015; Panagos et al. 2015b; Teng et al. 2016, 2018; Yang et al. 2003). Soil erosion models facilitate land management and in understanding sediment transport and its impact on the landscape (Benavidez et al. 2018).

Recently, the application of remote sensing (RS) and geographic information systems (GIS) with the RUSLE model in the soil erosion study became inevitable worldwide (Ganasri and Ramesh 2016; Gaubi et al. 2017; Gelagay 2016; Lee et al. 2017; Mukanov et al. 2019; Nasir and Selvakumar 2018; Nyesheja et al. 2018; Ostovari et al. 2017; Prasannakumar et al. 2012; Sujatha and Sridhar 2018; Thomas et al. 2018a, b). For example, the RUSLE C factor was estimated using NDVI and ground data and by calculating the exponential regression for mountain pastures in the southwest of Kyrgyzstan (Kulikov et al. 2016). Therefore, such methods can also be successfully applied to Central Asia.

It should be noted that in recent years, the climatic conditions in the Central Asian countries have changed due to the reduction of glacier areas of most of the Tien Shan (Aizen et al. 2007; Duishonakunov et al. 2014; Kenzhebaev et al. 2017) and Pamir-Alay (Chevallier et al. 2014; Hagg et al. 2007) mountain systems from the south and the drying of the Aral Sea (Issanova et al. 2018; Lioubimtseva and Henebry 2009) in the north. In this regard, the shortage of water for irrigation, especially of pastures, is felt; natural vegetation cover is degraded, erosion processes and salinization are increasing, and the productive capacity of irrigated lands is decreasing (Hamidov et al. 2016). Humanity is facing a serious problem – the preservation of existing natural landscapes, which includes improving and multiplying its types.

Rainfall erosivity is associated with the powerful kinetic energy of raindrops, which often separate soil elements and transport them along with surface runoff (Amanambu et al. 2019). To describe the processes of erosion, rainfall erosivity is the most significant factor and offers conservation actions by the models of soil erosion prediction (Panagos et al. 2017b). Recently, there is an emerging evidence of the influence of climate change on rainfall erosivity in various parts of the globe (Almagro et al. 2017; Amanambu et al. 2019; Gu et al. 2018; Gupta and Kumar

2017; Lai et al. 2016; Meusburger et al. 2012; Panagos et al. 2017a; Plangoen et al. 2013).

Global climate change has taken a permanent place among the significant environmental problems being faced by the world community in recent decades (Ibatullin et al. 2009). It is becoming one of the most significant environmental issues of the twenty-first century (Campbell et al. 2011; Carter et al. 2015). The temperature predictions for RCP4.5 show general warming in the Central Asian region close to 2 °C (2021–2050 compared with 1961–1990) by the CMIP5, and the heating tendency is more intensified for RCP8.5 (Immerzeel et al. 2013). Climate change is expected to have a significant impact on the Central Asian region based on global climate predictions (Unger-Shayesteh et al. 2013).

The modeling method is an ordinary and necessary way to predict the future effects of climate change. Simulation studies are often based on erosion models and global climate models (GCM). GCMs are commonly used to predict climatic characteristics under hypothetical or large-scale climate scenarios. Thus, these meteorological data for different scenarios act as input data for an erosion model to obtain climate-altered variables associated with soil erosion.

Remote sensing and GIS methods are often used to obtain other necessary input data for models. Several models with reliable results have been used in many forecasting or exploitation tasks or for planning. However, these models usually operate according to the quantity and quality of the available data, which are unevenly distributed. In developed countries, data are highly available, but not in developing countries, such as the former Soviet republics in Central Asia (Gafurov 2010). Usually, data availability is minimal in a high-mountain region or in less-populated areas, where continuous observations with daily time steps require incredibly high costs and labor, which is not feasible for most developing countries. Central Asia is one of those regions where data availability is minimal.

According to Kulikov (2018), for decades, Central Asia has remained a white spot on the map of international science because of two reasons. First, since most of the earlier scientific research was conducted in Soviet times, it was published in the Russian language, and since the Soviet Union was a closed country, the publications barely reached the English-speaking world. The second reason is because Central Asia does not have sufficient resources to conduct large-scale systematic baseline environmental studies.

Despite the importance and demand for studies on soil, vegetation, and climate, the unfortunate economic situation in the region does not allow for a regular assessment of their condition (Kulikov 2018). However, the modern developments in the field of GIS and RS and the availability of spatial data, together with the increase in computing power, make this task less expensive. Fieldwork and ground data collection, along with laboratory work, are the most time-consuming parts of the study, whereas auxiliary informational and analytical tools are now less significant due to the generosity of their excellent suppliers. This allows the developing countries to conduct a thorough assessment and regular monitoring of natural resources and hazards.

Chapter 2
Natural Conditions in Central Asia

Abstract This chapter provides information about the topography, climate conditions, and description about soils of Central Asia. Central Asia occupies a vast territory on the Asian continent, including the Kyrgyz Republic, the Republic of Tajikistan, the Republic of Turkmenistan, the Republic of Uzbekistan, and the Republic of Kazakhstan entirely, and covers an area of ~4 million km². The region is a separate natural-historical region, sharply differing. Significant differences in the height of parts of this territory – from the areas lying below the ocean level to the highest mountain peaks – make here variety of climate and landscape forms. Having a long and orographically complex territory with extensive lowlands and highest mountain elevations in the south, southeast, Central Asia is characterized by a variety of climatic conditions. The climate of Central Asia is distinguished by a high continentality, marked by a great amplitude of fluctuations in air temperature and a meagre amount of precipitation from adjacent areas by its natural conditions. Xerosols and Yermosols, followed by Kastanozems, Solonetz, and Lithosols with rock outcrops, occupy the largest area in Central Asia.

Keywords Central Asia · Climatic conditions · Tien-Shan · Soil · Topography · Natural conditions

2.1 Topography and Climatic Conditions in Central Asia

Central Asia occupies a vast territory in the Asian continent and includes the Kyrgyz Republic, the Republic of Tajikistan, Turkmenistan, Uzbekistan, and Kazakhstan (Alamanov et al. 2006) (Table 2.1). These five countries cover an area of 4 million km² (46°45'28.13"–87°21'47.81" E, 35°5'2.24"–52°33'30.49" N) (Chen and Zhou 2015). The nations in Central Asia were once part of the Soviet Union. In the physical–geographical terms, the region is a separate natural–historical region and is significantly different from the adjacent areas in terms of its natural conditions.

© The Author(s), under exclusive license to Springer Nature Switzerland AG 2021
E. Duulatov et al., *Current and Future Trends of Rainfall Erosivity and Soil Erosion in Central Asia*, SpringerBriefs in Environmental Science, https://doi.org/10.1007/978-3-030-63509-1_2

Table 2.1 Land resources and various agriculture indicators of Central Asian countries. After Nurbekov et al. (2016)

Country	Territory (M ha)	Cropland (M ha)	Irrigated land (M ha)	Rainfed land (M ha)	Per capita cropland (ha)	Agric. GDP (%)
Kazakhstan	272.5	24	1.6	22.1	1.41	5.3
Kyrgyzstan	20.0	1.4	1.1	0.3	0.23	25.8
Tajikistan	14.2	0.9	0.7	0.2	0.10	19.8
Turkmenistan	48.8	1.8	1.8	0.0	0.33	22.1
Uzbekistan	44.7	4.3	4.3	0.5	0.15	19.4
Total	**400.3**	**33**	**9.5**	**23.1**	**0.48**	**9.9**

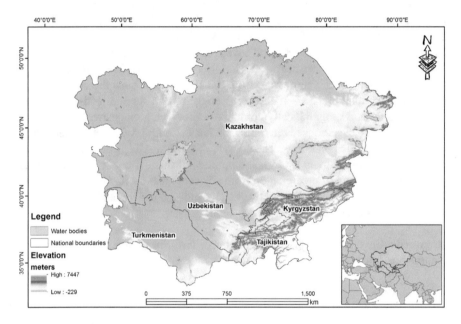

Fig. 2.1 Digital Elevation Model of Central Asia, 90-m Shuttle Radar Topography Mission Digital Elevation Model (SRTM DEM 90 m (Jarvis et al. 2008))

The local Central Asian climates are of three types: (1) climates of the temperate zone, (2) climates of the dry subtropical zone, and (3) mountain climates of Tien Shan, Pamir-Alai, Pamir, and Kopetdag with a well-marked altitudinal-belt zonality (Alamanov et al. 2006). The elevation ranges from −229 to 7447 m (Fig. 2.1). The average annual precipitation in Central Asia is 254 mm, the minimum is 66 mm, and the maximum is 1222 mm (Fig. 2.2).

The mountainous regions in Central Asia represent the area of feeding and the formation of river flows and are characterized by a well-developed river network. A characteristic feature of the region is the transboundary nature of the overwhelming majority of rivers, which directly impacts the formation of the water policy of the

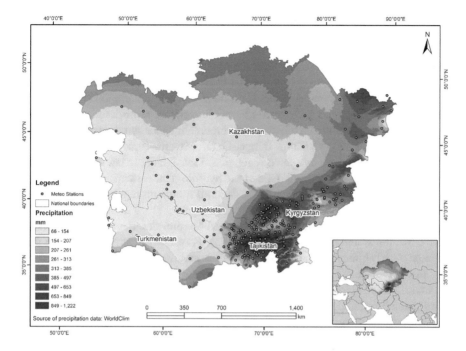

Fig. 2.2 Precipitation in Central Asia. (Source: WorldClim data (Hijmans et al. 2005))

countries located within it. The basin of the largest river in the Amu Darya region belongs to Tajikistan, Kyrgyzstan, Afghanistan, Iran, Uzbekistan, and Turkmenistan.

Syr Darya flows through the territories of Kyrgyzstan, Uzbekistan, and Kazakhstan. The main tributaries of the Chinese Tarim River flow from the mountains of Kyrgyzstan and Tajikistan. Irtysh and Ili, the flow of which has a great economic and environmental importance for Kazakhstan and Russia, originate in China.

Dozens of small- and medium-sized rivers flowing into the Fergana part of Uzbekistan and Tajikistan are formed on the slopes of the Kyrgyz mountains – the Fergana, Turkestan, and Alai ranges. The differences in the interests of water management among the countries of the upper river basins and their lower reaches largely determine the difficulties in the development and implementation of a single optimized water policy in the region.

The relief of Central Asia is high in the east but low in the west, which is mostly dominated by plains and hills (Zhang et al. 2018). The central part of the region is occupied by the Turan Lowland, whose height does not exceed 100–200 m above sea level. The Kazakh Lowland is located in the north, and its height rarely exceeds 300–400 m and reaches 1000 m in only some places.

The Ustyurt Plateau extends between the Caspian and Aral seas and ends in all directions with a steep ledge called Chink. From the south and east, the Turan

Lowland is bordered by the Pamir and Tien Shan mountain ranges, with heights of up to 7495 m (Somoni Peak, former Communism), and Kopet Dag, with heights of up to 2000–3000 m. The northern part of the region known as Central Kazakhstan is a semi-desert area. Here, along with steppe feather grass meadows, there are areas of bare soil devoid of any vegetation (Alamanov et al. 2006).

Hydrologically, a scarce river network, mostly drying up temporary streams, an abundance of saline, and self-flowing lakes characterize semi-deserts. For Central Asia, vast deserts are characteristic, with mountain systems bordering it from all sides: Kara-Kum, Kyzyl-Kum, Betpak-Dala, Moyun-Kum, and Takla-Makan. Deserts are characterized by sparse vegetation. Transit large rivers – Amudarya, Syr Darya, and Tarim – are characteristic of the hydrological network. In the mountainous areas of the Pamirs and Tien Shan, as in all mountain regions, a high-altitude zoning of natural conditions exists.

The foothills of the mountains are occupied by semi-deserts, which are replaced above by steppe and forest-steppe landscapes. Forests form above them, and then subalpine and alpine meadows. The ridge of the mountain ranges is a rocky surface, and in some places, it is covered with eternal snow and glaciers. The mountains in Central Asia are among the regions of the globe with the most significantly developed glaciation. In Central Asia, the total number of glaciers reaches 17,000, and the area of glaciation covers more than 17,000 km^2. The largest glacier in the Pamirs is the Fedchenko Glacier, which is 71-km long, and the South Engilchek Glacier in Tien Shan, which is 60-km long. The snow frontier is at the level of 3000–3600 m in the peripheral mountain ranges and 5500 m in the central parts of the Pamir mountain ranges (Alamanov et al. 2006).

The significant differences in the height of the parts of this territory – from the areas lying below the ocean to the highest mountain peaks – make here an extraordinary variety of surface shapes, climate and landscape. Having a lengthy and orographically complicated territory with extensive lowlands and high mountain elevations in the south and southeast, Central Asia is characterized by a variety of local climatic conditions.

The climate in Central Asia has a local feature, namely, high continentality, which is characterized by a large amplitude of fluctuations in air temperature and small precipitation (Alamanov et al. 2006). All the three types of climates mentioned above are formed due to the incoming solar radiation, physiographic conditions of the territory, and circulation of the atmosphere. Along with the circulation of the atmosphere and radiation factors, its physiographic position in the center of the largest continent – Eurasia and the surface structure – plays a significant role in the vast plains and mountains in the south.

In most areas, the climate is dry and continental, with hot summers and cold winters with occasional snowfalls. Outside the highlands, the climate is mostly semi-arid to dry. In lower altitudes, summer is hot and accompanied by bright sunshine. In winter, it sometimes rains and snows from low-pressure systems that cross an area from the Mediterranean. In most of the region, precipitation is at its maximum in spring, which is associated with the migration to the north of the Iranian branch of the Polar Front (Lioubimtseva and Henebry 2009).

The average monthly rainfall is low from July to September, increases in autumn (October and November), and is thus maximum in March or April, after which a rapid drying occurs in May and June. Winds can be intense, which sometimes cause dust storms, especially towards the end of the dry season in September and October. Specific cities that exemplify climate patterns in Central Asia include Tashkent and Samarkand, Uzbekistan; Ashgabat, Turkmenistan; and Dushanbe, Tajikistan, the latter of which represents one of the wettest climates in Central Asia with an average annual rainfall of over 254 mm (Alamanov et al. 2006).

2.2 Description of Central Asian Soils

The natural conditions in Central Asia are very diverse. Depending on the climate, vegetation, and geological structure of the area, the soil cover also changes. The most pronounced change in soil types occurs from north to south, which is called latitudinal zoning. In the mountains, the soil types change from the foot to the top, indicating that there is a vertical (high-altitude) zonation.

The weathering processes in the sharply continental conditions in Central Asia led to the formation of a residual carbonate–silicate weathering crust, which is relatively weakly weathered, that is expressed in the abundant accumulation of sandy–silty destruction products and low content of fine (silty) particles. The weathering products are loess-like (this is one of the main reasons of the wide distribution of loess and loess-like deposits in Central Asia); they are rich in various elements of plant ash nutrition.

The weathering processes are most active in the middle-mountain and low-mountain high-altitude zones, where more clayey products are formed. In deserts, however, weathering is mostly weakened and complicated by the accumulation of carbonates, gypsum, and readily soluble salts. Similar processes occur in the desert highlands.

The products of weathering of rocks are transported by surface and underground runoff. Due to the drainlessness of the territory of Central Asia, they accumulate in closed basins. The main role in this accumulation belongs to readily soluble chloride-sulfate salts, gypsum and calcium carbonate. (Gvozdetsky and Mikhailov 1978).

All the desert plains in Central Asia are areas of chloride–sulfate accumulation, which occurs against a highly calcified background since solid runoff also plays a significant role in the accumulation of lime (suspended sandy, dusty, and clayey particles contain 7–23% of carbon dioxide). In addition, the low desert plains in Central Asia, which have experienced several stages of the alluvial regime, are areas where aluminosilicate weathering products accumulate in the form of sandy, sandy loam, and clay particles brought by surface waters from the mountain ranges.

The composition of rocks and the processes of alluvial and proluvial sedimentation and subaerial denudation, as well as glaciation, gravitational processes, and plane washout in the mountains, have a great influence on weathering and soil formation.

Artificial irrigation is an important factor in soil formation. Long-term irrigation had a beneficial effect on the soils of the oases, but led to salinization and waterlogging of the surrounding lands. The destruction of natural vegetation in deserts caused soil deflation in sandy areas, and the destruction of trees and shrubs in the mountains and foothills led to increased erosion and steppe formation of forest soils.

The climatic conditions of soil formation on the plains of Central Asia change from north to south. In this regard, the steppe chernozem soil zone of the far north of Central Kazakhstan is replaced to the south by a zone of chestnut soils and then a zone of brown desert-steppe soils.

The desert plains in Central Asia belong to the zones of gray-brown soils (northern and southern), where, by the diversity of parent rocks, various soils are distinguished, which belong to the types of gray-brown soils, takyr soils, and takyrs. In the southern half of the desert zone, in ephemeral deserts, and in the semi-deserts of the foothill plains and foothills, typical gray soils are widespread (Gvozdetsky and Mikhailov 1978).

In the mountains in Central Asia, the following altitude zones are distinguished (Gvozdetsky and Mikhailov 1978):

1) desert steppe, with the corresponding subtypes of brown and gray earth soils;.
2) Dry steppe with mountain chestnut, dark gray, and brown soils;
3) forest-meadow-steppe with various soils – mountain-brown, mountain-forest brown, dark-brown and dark-colored, mountain chernozems;
4) mountain-meadow and meadow-steppe soils with different subtypes of subalpine mountain-meadow and meadow-steppe soils;
5) high-mountain meadow, meadow-steppe, desert-steppe, desert takyr-like, and dry tundra soils;
6) zone of perennial snows and glaciers.

In the altitudinal zonation of the soil cover, there are large differences between the northern and southern ridges, between better- and worse-moisturized parts. The altitude boundaries of these zones are also not the same. These determine the existence of different spectra of altitudinal zoning. The zones are not equally well expressed everywhere; some of them fall out in places or are represented by isolated areas, such as the forest-meadow-steppe zone. Certain types of soils in the highland zone have signs of soil-forming processes, which are more common in the mountains in Central Asia.

Some soils of the desert zone, especially typical gray soils, are very fertile under the conditions of artificial irrigation and constitute the main base of oasis agriculture, mainly cotton growing. Other desert soils are used for irrigated agriculture after appropriate reclamation. In the desert zone, there are large areas of virgin lands that are suitable for irrigated agriculture. Typical and dark-gray soils and some other soils of the high-elevated foothill plains and foothills are used in rainfed agriculture. During the development of virgin and fallow lands in Kazakhstan, large tracts of land were plowed up in the zones of chernozem and chestnut soils.

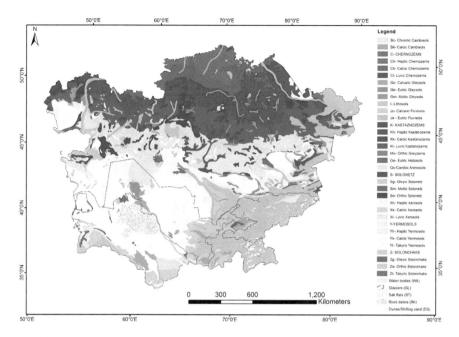

Fig. 2.3 Soil map of Central Asia (FAO DSMW)

Xerosols and Yermosols, and then Kastanozems, Solonetz, and Lithosols with rock outcrops, occupy the largest area in Central Asia (Fig. 2.3). Sand dunes prevail in the Kyzyl-Kum and Kara-Kum deserts in the south (Sommer and de Pauw 2011).

Chapter 3
Data Sources and Methodology

Abstract This chapter presents the data available together with their analysis for their suitability for the intended research. Projected rainfall data from GSMs BCCCSM1-1, IPSLCM5ALR, MIROC5, and MPIESMLR (Central Asia); and GISSE2H, HadGEM2ES, and NorESM1M (Kazakhstan) for the RCP2.6 and RCP8.5 greenhouse emission scenarios were used. This chapter analyzes a detailed methodology on the RUSLE model, predicting the impact of climate change on rainfall erosivity and soil erosion based on geographic information system GIS and remote sensing (RS) techniques to assess soil erosion. The performance of the rainfall erosivity model was assessed by comparing the rainfall erosivity of observation data with that of the baseline data using coefficient of determination (R^2), root-mean-square error (RMSE), and Nash–Sutcliff efficiency (NSE).

Keywords GCMs · RCPs · WorldClim · Rainfall erosivity · Soil erosion · Soil erodibility · Cover management · Support practice · Slope length and steepness · Erosivity density · Central Asia · Climate change · Delta method · RUSLE · GIS · RS · Mann-Kendall trend test

3.1 Projected and Observed Rainfall Data

Compared with the Coupled Model Intercomparison Project Phase 3 (CMIP3), CMIP5 is a remarkable improvement as it uses a new set of emission scenarios called RCPs (Amanambu et al. 2019; Taylor et al. 2012). The projected rainfall data obtained from GSMs BCCCSM1-1, IPSLCM5ALR, MIROC5, and MPIESMLR (Central Asia) and GISSE2H, HadGEM2ES, and NorESM1M (Kazakhstan) for the RCP2.6 and RCP8.5 greenhouse emission scenarios were used (Table 3.1) (Taylor et al. 2012).

The GCMs were selected owing to their relative independence and good performance in the precipitation simulation for Central Asia (Luo et al. 2019) and the

© The Author(s), under exclusive license to Springer Nature Switzerland
AG 2021
E. Duulatov et al., *Current and Future Trends of Rainfall Erosivity and Soil Erosion in Central Asia*, SpringerBriefs in Environmental Science,
https://doi.org/10.1007/978-3-030-63509-1_3

17

Table 3.1 Global Climate Models (GSMs) from the Climate Change Agriculture and Food Security (CCAFS, http://www.ccafs-climate.org) portal

Model	Institute	Country	Resolution
Central Asia			
BCCCSM-1.1	Beijing Climate Center, Climate System Model 1.1	China	~2.8125° × 2.8125°
IPSLCM5BLR	Institute Pierre Simon Laplace Model, New Atmospheric Physic at Low Resolution	France	3.75° × ~1.9°
MIROC-5	Model for Interdisciplinary Research On Climate	Japan	1.4° × 1.4°
MPIESMLR	Max Planck Institute for Meteorology	Germany	1.875° × ~1.9°
Kazakhstan			
GISSE2H	Goddard Institute for Space Studies, ModelE2/Hycom	US	2° × 2.5°
HadGEM2-ES	Hadley Centre Global Environmental Model 2 – Earth System	UK	2.5° × 2°
NorESM1M	Norwegian Climate Centre	Norway	2.5° × ~1.9°

Tibetan Plateau (Teng et al. 2018). Global precipitation with ~1-km^2 horizontal resolution was obtained from the WorldClim database (Hijmans et al. 2005). In assessing the future changes in the erosion of rainfall and possible consequences, the predicted rainfall data for the "near" (2020–2049) and "far" future (2060–2089) have been downloaded from the Climate Change Agriculture and Food Security portal (CCAFS, http://www.ccafs-climate.org).

The data were statistically downscaled to ~1-km^2 horizontal resolution using the delta method (Ramirez-Villegas and Jarvis 2010), based on the sum of interpolated anomalies to high-resolution monthly climate surfaces from WorldClim (Hijmans et al. 2005). These anomalies were then interpolated using thin-plate spline interpolation (Ramirez-Villegas and Jarvis 2010). Moreover, these datasets were used as input data for this study.

To evaluate the precipitation of the baseline output, we used the precipitation data from 150 rain gauge stations obtained from the Kazakhstan Hydro and Meteorological Agency (Kazhydromet). Also, the "Central Asia Temperature and Precipitation (CATP) data (1879–2003), Version 1" data, obtained from the National Snow and Ice Data Center (NSIDC) (Williams and Konovalov 2008). The data of 270 rain gauges provided by the Hydrometeorological Agencies of Central Asian countries. This dataset contains monthly climatic data.

3.2 RUSLE Model and Its Factors

The RUSLE parameters were calculated using separate equations with input data from rainfall, satellite images, and DEM. The source data, their sources, and the equations used are presented in Table 3.2. The equations available in the literature

Table 3.2 Data sources and their descriptions

Data	Source	Resolution
Precipitation	Kazakhstan Hydrometeorological Agency (Kazhydromet)/Central Asia Temperature and Precipitation Data (Williams and Konovalov 2008)/WorldClim (Hijmans et al. 2005)	Monthly/1 km
DEM	Shuttle Radar Topography Mission (SRTM) (Jarvis et al. 2008)	90 m
Soil	Food and Agriculture Organization (FAO) Harmonized World Soil Data (HWSD) (Nachtergaele et al. 2010)	1 km
NDVI	16 days Moderate Resolution Imaging Spectroradiometer (MODIS) from the National Aeronautics and Space Administration (NASA)	250 m
LULC	European Space Agency Climate Change Initiative Land Cover (ESACCI-LC, http://www.maps.elie.ucl.ac.be/)	300 m

for calculating the factors have been tested iteratively, and the optimal equations are selected based on their suitability for use with the available data and the ability to produce estimates comparable with the published field erosion measurements. The calculation of individual factors is described in more detail in the following subsubsections.

RUSLE is the technique most extensively used globally to predict the long-term rate of erosion. Wischmeier and Smith (1965) from the US Department of Agriculture first developed the Universal Soil Loss Equation (USLE) as a field-scale model (Wall et al. 2002). In 1997, it was revised to better assess the values of different factors in USLE (Renard et al. 1997). Since then, the RUSLE model has been well studied and extensively used to estimate soil erosion in the areas under consideration at different scales (Ganasri and Ramesh 2016; Karamage et al. 2016; Lee et al. 2017; Mukanov et al. 2019; Nyesheja et al. 2018; Prasannakumar et al. 2012).

$$A = R \times K \times LS \times C \times P \tag{3.1}$$

where A denotes the calculated average loss of soil per unit of area, expressed in t ha^{-1} year^{-1}. R denotes rainfall–runoff erosivity (MJ mm ha^{-1} h^{-1} year^{-1}). K is the soil erodibility factor reflecting the susceptibility of soil to erosion (t h MJ^{-1} mm^{-1}). LS is the topographic factor, which includes the slope length (L) and the slope steepness (S) factors. C is the cover and management factor. P is the support and conservation practice factor (LS, C, and P factors are unitless).

3.2.1   Rainfall–Runoff Erosivity (R) Factor

In this study, the rainfall erosivity (R) factor from the RUSLE model was selected to estimate the changes in rainfall erosivity. Rainfall erosivity was calculated using the precipitation values of gridded GCMs, comparing it with the WorldClim data. Wischmeier and Smith (1978) and Renard et al. (1997) described the original method of calculating erosivity as follows:

$$R = \frac{1}{n}\sum_{j=1}^{n}\sum_{k=1}^{m_j}(EI_{30})k, \qquad (3.2)$$

where R is the mean annual rainfall erosivity (MJ mm ha^{-1} h^{-1} year^{-1}), n is the number of years of data, m_j is the number of erosive events in the j year, and EI_{30} is the rainfall erosivity index of a storm k. The event's rainfall erosivity index EI_{30} is defined as follows:

$$EI_{30} = I_{30}\left(\sum_{r=1}^{m}e_r v_r\right) \qquad (3.3)$$

where e_r is the unit rainfall energy (MJ ha^{-1}), and v_r is the rainfall depth (mm) during a time period r. I_{30} is the maximum rainfall intensity during a 30-min period of the rainfall event (mm h^{-1}).

$$e_r = 0.29\left[1 - .072\exp\left(-0.05i_r\right)\right] \qquad (3.4)$$

where i_r is the rainfall intensity during the period (mm h^{-1}).

The information required to calculate the R factor using the proposed method is usually difficult to obtain in many parts of the world. Therefore, various studies have been conducted to derive regression equations for the derivation of the R factor. These simplified methods provide great convenience for studying the spatial and temporal variabilities of rainfall erosivity. Renard and Freimund (1994) proposed the following equations for estimating the R factor using annual precipitation or the Modified Fournier Index (MFI) when there are no data on rainfall intensity for a particular site:

$$R = 0.04830 \times P^{1.61}, \text{where } P < 850mm \qquad (3.5)$$

$$R = 587.8 - 1.219 \times P^2, \text{where } P \geq 850mm \qquad (3.6)$$

where R is the rainfall erosivity factor (MJ mm ha^{-1} h^{-1} year^{-1}), and P is the average mean annual precipitation.

$$R = 0.7397MFI^{1.847}, \text{where } MFI < 55mm \qquad (3.7)$$

$$R = 95.77 - 6.081MFI + 0.4770MFI^2, \text{where } MFI \geq 55mm \qquad (3.8)$$

where R is the rainfall erosivity (MJ mm ha^{-1} h^{-1} year^{-1}). MFI is the Modified Fournier Index given below (Arnoldus 1977; Arnoldus 1980):

$$MFI = \sum_{i=1}^{12}\frac{p_i}{P} \qquad (3.9)$$

where P is the annual precipitation (mm), and p_i is the monthly rainfall.

In this study, the rainfall erosivity has been determined using the average annual precipitation (Eqs. 3.5 and 3.6). We used these equations because they have been widely used in similar studies (Amanambu et al. 2019; Mukanov et al. 2019; Yang et al. 2003). The data used to derive the R factor are gridded WorldClim data of precipitation and the GCMs.

3.2.2   Soil Erodibility Factor (K)

The K factor quantitatively determines the cohesive nature of the soil type, which depends on the physical and chemical properties of the soil, thereby contributing to its erodibility. These numerical indicator actions the vulnerability of soil to erosion rainfall impacts (Renard et al. 1997; Wischmeier and Smith 1978). The main soil properties affecting the K factor are soil texture, organic matter, soil structure, and soil profile permeability (Uddin et al. 2016). Soil data were obtained from the Food and Agriculture Organization, the Harmonized World Soil Data (FAO HWSD) (Nachtergaele et al. 2010; Nachtergaele et al. 2009). The soil erodibility affects the transportability of sediments and the vulnerability of the soil surface to erosion. Equation (3.10), proposed by Yang et al. (2003), is expressed as follows:

$$K = \frac{1}{7.6}\left\{0.2 + 0.3\exp\left[-0.0256SAN\left(1 - \frac{SIL}{100}\right)\right]\right\}\left(\frac{SIL}{CLA + SIL}\right)^{0.3}$$

$$\left(1 - \frac{0.25OM}{OM + \exp(3.72 - 2.95OM)}\right) \tag{3.10}$$

$$\left(1 - \frac{0.75SN}{SN + \exp(-5.51 + 22.9SN)}\right)$$

where $SN = 1 - SAN/100$, and SAN % is the content of sand, SIL % is the content of silt, CLA % is the content of clay, and OM is % the content of organic matter.

3.2.3   Slope Length and Steepness (LS) Factor

The topographic (LS) factor is the product of the coefficient of the slope length (L) and steepness (S). The LS factor layer resulted from the Shuttle Radar Topography Mission (SRTM) Digital Elevation Model (DEM) 90 m, which was obtained from the USGS Earth Explorer database (Table 3.3). The L factor was calculated using the methodology described by Desmet and Govers (1996).

Table 3.3 P values by different land use/land cover (LULC) types

Land cover type	%	Area, km²	P factor
Cropland rainfed	0.4	886.2	0.5
Herbaceous cover	1.9	3705.5	1
Cropland irrigated or post-flooding	15.1	30239.6	0.5
Mosaic cropland (>50%)/natural vegetation (tree shrub) herbaceous cover) (<50%)	2.4	4851.8	0.8
Mosaic natural vegetation (tree shrub herbaceous cover) (>50%)/ cropland (<50%)	5.5	10905.2	0.8
Tree cover broadleaved deciduous closed to open (>15%)	0.1	227.8	1
Tree cover broadleaved deciduous closed (>40%)	0.0	0.1	1
Tree cover needleleaved evergreen closed to open (>15%)	1.6	3130.4	1
Tree cover needleleaved evergreen closed (>40%)	0.0	1.5	1
Tree cover needleleaved evergreen open (15–40%)	0.0	0.9	1
Tree cover needleleaved deciduous closed to open (>15%)	2.8	5581.6	1
Tree cover needleleaved deciduous closed (>40%)	0.0	0.6	1
Tree cover mixed leaf type (broadleaved and needleleaved)	0.0	2.4	1
Mosaic tree and shrub (>50%)/herbaceous cover (<50%)	2.0	3985.7	1
Mosaic herbaceous cover (>50%)/tree and shrub (<50%)	2.0	4071.3	1
Shrubland	1.5	3052.8	1
Grassland	38.0	76011.4	1
Sparse vegetation (tree shrub herbaceous cover) (<15%)	10.6	21238.6	1
Urban areas	0.3	651.2	1
Bare areas	9.0	17942.5	1
Consolidated bare areas	0.0	0.6	1
Unconsolidated bare areas	0.0	5.4	1
Water bodies	3.6	7285.0	0
Permanent snow and ice	3.0	6028.4	0

$$L_{i,j} = \frac{\left(A_{i,j-in} + D^2\right)^{m+1} - A_{i,j-in}^{m+1}}{D^{m+2} * x_{i.j}^m * 22.13^m} \tag{3.11}$$

where $L_{i,j}$ is the slope length factor from the grid cell with coordinates (i, j), $A_{i,j-in}$ is the contributing area at the inlet of the grid cell measured in m², D is the grid cell size, in m, $x_{i,j} = (\sin\alpha_{i,j} + \cos\alpha_{i,j})$, $\alpha_{i,j}$ is the aspect direction for the grid cell with coordinates (i, j), and m is related to the ratio β of the rill to interill erosion:

$$m = \frac{\beta}{(1+\beta)} \tag{3.12}$$

$$\beta = \frac{\left(\sin\theta / 0.0896\right)}{\left[3*\left(\sin\theta\right)^{0.8} + 0.56\right]} \tag{3.13}$$

$$S = \begin{cases} 10.8\sin\theta + 0.03, \theta < 9\% \\ 16.8\sin\theta - 0.5, \theta \geq 9\% \end{cases} \tag{3.14}$$

where θ is the slope angle.

For very steep slopes which are not correctly covered by the S factor, Schmidt et al. (2019) developed the S factor equation for the alpine environment (Eq. 3.15):

$$S_{\text{alpine}} = 0.0005s^2 + 0.7956s - 0.4418 \tag{3.15}$$

where s is the slope steepness in percent.

In this work, the LS factor was implemented in the System for Automated Geoscientific Analyses (SAGA) 7.1.1 software. The result of the LS values was mapped in ArcGIS 10.2.

3.2.4   Cover Management (C) Factor

The C factor compares the loss of soil from an exact type of vegetation cover. A widely used approach to the C factor estimation is based on the use of remote sensing data (Borrelli et al. 2018; Kulikov et al. 2016; Panagos et al. 2015a; Schmidt et al. 2018). The *C* factor was assessed using the MODIS NDVI (June–August 2017). MODIS 16-days indicates NDVI 250-m resolution, obtained from the National Aeronautics and Space Administration (NASA). NDVI is described in Eq. (3.16) (Tucker 1978).

$$NDVI = \frac{NIR - Red}{NIR + Red} \tag{3.16}$$

where *Red* is the visible red, and *NIR* is the near-infrared spectral reflectance measurements. The values of NDVI range from −1 to 1. Equation (3.17) was used to produce the C factor (Almagro et al. 2019). The C factor was proposed by Durigon et al. (2014) and adapted by Colman (2018).

$$C = 0.1\left(\frac{-NDVI + 1}{2}\right) \tag{3.17}$$

3.2.5 *Conservation Practice (P) Factor*

The conservation practice factor (P) was derived based on the land cover map of the study area (Table 3.3). The conservation practice factor is the relation of soil losses by an exact practice of crop support, compared with the attitude of straight-line agriculture up and down the slope. Land use and land cover data (2010), derived from the European Space Agency Climate Change Initiative Land Cover (ECACCI-LC) data center. P factor derived from Yang et al. (2003).

3.3 Annual Erosivity Density Ratio

According to Kinnell (2010), the erosivity density coefficient is the ratio of rainfall erosivity (R) factor to precipitation. In practice, it measures the erosivity per unit of precipitation (mm) and is expressed as MJ ha^{-1} h^{-1}.

$$ED = \frac{R}{P} \tag{3.18}$$

where ED is the erosivity density, R is the average annual rainfall erosivity, and P is the average annual precipitation.

3.4 Trend Test

The Mann–Kendall (MK) nonparametric test was used to test for trends in Kazakhstan. The purpose of the MK test (Kendall 1975; Mann 1945) was to statistically assess the presence of a monotonous upward or downward trend of the variable of interest over time. Annual data series are used to analyze the trends in this study. The MK test was used in different countries to test the trends in rainfall erosivity (Gedefaw et al. 2018; Gu et al. 2018; Lai et al. 2016; Sadeghi and Hazbavi 2015). The data of individual time series of rainfall erosivity are compared with all the corresponding data of the time series of the year. Thus, the MK test statistics is the cumulative result of all data values. Statistics test Mann–Kendall S is equal to Eq. (3.19)

$$S = \sum_{i=1}^{n-1} \cdot \sum_{j=i+1}^{n} \operatorname{sgn}\left(Y_i - Y_j\right) \tag{3.19}$$

The trend test is applied to Y_i data values ($i = 1, 2, \ldots n - 1$) and Y_j ($j = i + 1, 2, \ldots n$). The data values of Y_i are used as reference points for comparison with the data values of Y_j which is expressed as follows:

$$\operatorname{sgn}\left(Y_i - Y_j\right) = \begin{cases} +1 \, if \left(Y_i - Y_j\right) > 0 \\ 0 \, if \left(Y_i - Y_j\right) = 0 \\ -1 \, if \left(Y_i - Y_j\right) < 0 \end{cases} \tag{3.20}$$

where Y_i and Y_j are values in periods i and j.

The variance $\operatorname{Var}(S)$ is given by

$$\operatorname{Var}(S) = n(n-1)(2n+) - \sum_{k=1}^{m} t_k \left(t_k - 1\right)\left(2t_k + 5\right) \tag{3.21}$$

where n is the number of rainfall erosivity events, m is the number of tied events, and t_k denotes the number of extent k. The MK test statistics based on the normal distribution is given as follows:

$$Z_s = \begin{cases} \dfrac{S-1}{\delta} \, if \, S > 0 \\ 0 \, if \, = 0 \\ \dfrac{S+1}{\delta} \, if \, S < 0 \end{cases} \tag{3.22}$$

When Z_s is greater than 0, it indicates an upward trend, and when Z_s is less than 0, it indicates a downward trend.

3.5   Model Evaluation Rainfall Erosivity

To evaluate the R factor of the baseline output, we used the precipitation data from the Central Asia Temperature and Precipitation (CATP) data (1879–2003), Version 1, obtained from the National Snow and Ice Data Center (NSIDC) (Williams and Konovalov 2008). Also, the precipitation data from 150 meteorological stations were obtained from the Hydrometeorological Agency of the Republic of Kazakhstan (Kazhydromet). This dataset contains monthly climatic data. The performance of the rainfall erosivity model was assessed by comparing the rainfall erosivity of the observation data with that of the baseline data using the correlation coefficient (r), root-mean-square error (RMSE), and Nash–Sutcliff efficiency (NSE) (Nash and Sutcliffe 1970) (Eqs. 3.23, 3.24, and 3.25 respectively).

$$R = \frac{\sum_{i=1}^{n}\left(Y_i^{obs} - Y_i^{mean}\right)\left(Y_i^{mod} - Y_i^{mean}\right)}{\sqrt{\sum_{i=1}^{n}\left(Y_i^{obs} - Y_i^{mean}\right)^2}\sqrt{\sum_{i=1}^{n}\left(Y_i^{mod} - Y_i^{mean}\right)^2}} \tag{3.23}$$

$$RMSE = \sqrt{\frac{\sum_{i=1}^{n}\left(Y_i^{obs} - Y_i^{mod}\right)^2}{n}} \tag{3.24}$$

$$NSE = 1 - \left[\frac{\sum_{i=1}^{n}\left(Y_i^{obs} - Y_i^{mod}\right)^2}{\sum_{i=1}^{n}\left(Y_i^{obs} - Y_i^{mean}\right)^2}\right] \tag{3.25}$$

where Y_i^{mod} is the baseline rainfall erosivity, Y_i^{obs} is the observed rainfall erosivity, and Y_i^{mean} is the mean of the observed and baseline rainfall erosivity.

Chapter 4
Projected Rainfall Erosivity and Soil Erosion in Central Asia

Abstract The GCMs (BCCCSM1-1, IPSLCM5BLR, MIROC5, and MPIESMLR) were statistically downscaled using the delta method under Representative Concentration Pathways (RCPs) 2.6 and 8.5 for two periods: "Near" and "Far" future (2030s and 2070s). These GCMs data were used to estimate rainfall erosivity and its projected changes over Central Asia. WorldClim data were used as the present baseline precipitation scenario for the study area. The rainfall erosivity (R) factor of the Revised Universal Soil Loss Equation (RUSLE) was used to determine rainfall erosivity. The results show an increase in the future periods of the annual rainfall erosivity compared to the baseline. For all GCMs (BCCCSM1-1, IPSLCM5BLR, MIROC5, and MPIESMLR), with an average change in rainfall erosivity of about 5.6% (424.49 MJ mm ha^{-1} h^{-1} year^{-1}) in 2030s and 9.6% (440.57 MJ mm ha^{-1} h^{-1} year^{-1}) in 2070s as compared to the baseline of 402 MJ mm ha^{-1} h^{-1} year^{-1}. The magnitude of the change varies with the GCMs, with the largest change being 26.6% (508.85 MJ mm ha^{-1} h^{-1} year^{-1}), occurring in the MIROC-5 RCP8.5 scenario in the 2070s. Although annual rainfall erosivity shows a steady increase, IPSLCM5ALR (both RCPs and periods) shows a decrease in the average erosivity. Higher rainfall amounts were the prime causes of increasing spatial-temporal rainfall erosivity.

Keywords Rainfall erosivity · Erosivity density · Soil erosion · Central Asia · Tien-Shan · Climate change · GCMs · RCPs · RUSLE · GIS · RS · Baseline · Future scenarios

4.1 RUSLE Model Factors

RUSLE model was used to calculate the potential annual soil loss in Central Asia. The input layers were generated using remote sensing data and integrated into GIS to assess potential soil loss. The potential soil erosion rate for each pixel was com-

© The Author(s), under exclusive license to Springer Nature Switzerland AG 2021
E. Duulatov et al., *Current and Future Trends of Rainfall Erosivity and Soil Erosion in Central Asia*, SpringerBriefs in Environmental Science, https://doi.org/10.1007/978-3-030-63509-1_4

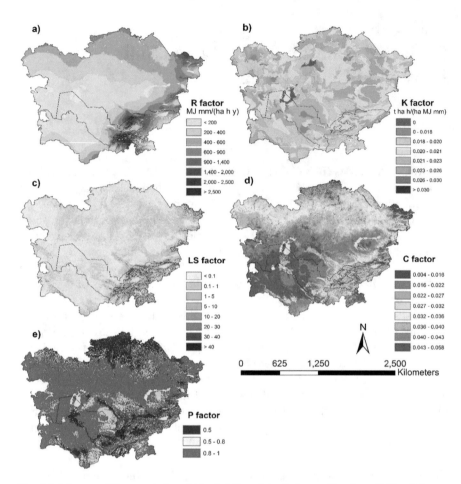

Fig. 4.1 Maps at a 250-m resolution of (**a**) rainfall erosivity, R factor; (**b**) soil erodibility, K factor; (**c**) slope length and steepness, LS factor; (**d**) cover management, C factor; and (**e**) support practice, P factor

puted by the integration of erosion factors described in the RUSLE model. These factors of soil erosion are described below:

The maps of the RUSLE factors in Central Asia are presented in Fig. 4.1. The average annual rainfall erosivity was calculated for the current period (1960–2000) using the WorldClim precipitation data. The mean rainfall erosivity was 870.3 MJ mm ha^{-1} h^{-1} year^{-1} with a standard deviation of 567.3 MJ mm ha^{-1} h^{-1} year^{-1} (Fig. 4.1 and Table 4.1). The highest annual R factor was observed in the southeastern part of Central Asia, with moderate values in the northern regions, but decreased westwards, where the lowest values were observed. However, the values gradually increased towards Tajikistan but decreased in the western parts occupied by Turkmenistan. The spatial distribution of the R factor constantly varied with regard to the annual precipitation in Central Asia. The estimated average annual rainfall

Table 4.1 The statistics related to RUSLE equation factors in Central Asian countries

Factors	Countries	Min	Max	Mean	STD
R	Kazakhstan	82.8	2915.5	370.0	229.3
	Kyrgyzstan	144.2	4409.0	870.0	566.9
	Tajikistan	41.1	4509.5	1444.1	1027.6
	Turkmenistan	78.9	1366.1	189.9	117.0
	Uzbekistan	77.6	3630.3	285.0	381.4
	Central Asia	**41.6**	**4509.5**	**401.5**	**405.0**
K	Kazakhstan	0.000	0.052	0.020	0.003
	Kyrgyzstan	0.000	0.041	0.020	0.002
	Tajikistan	0.018	0.038	0.020	0.001
	Turkmenistan	0.000	0.024	0.017	0.001
	Uzbekistan	0.000	0.021	0.019	0.005
	Central Asia	**0.000**	**0.050**	**0.020**	**0.000**
LS	Kazakhstan	0.0	496.5	1.5	8.9
	Kyrgyzstan	0.0	487.0	29.3	44.5
	Tajikistan	0.0	525.8	37.7	53.5
	Turkmenistan	0.0	372.1	1.0	5.5
	Uzbekistan	0.0	444.7	2.7	14.0
	Central Asia	**0.0**	**525.8**	**4.2**	**18.8**
C	Kazakhstan	0.005	0.058	0.035	0.008
	Kyrgyzstan	0.006	0.055	0.031	0.011
	Tajikistan	0.009	0.057	0.040	0.009
	Turkmenistan	0.008	0.058	0.043	0.004
	Uzbekistan	0.006	0.058	0.041	0.006
	Central Asia	**0.005**	**0.058**	**0.040**	**0.008**
P	Kazakhstan	0.50	1.00	0.90	0.18
	Kyrgyzstan	0.50	1.00	0.89	0.19
	Tajikistan	0.50	1.00	0.90	0.19
	Turkmenistan	0.50	1.00	0.93	0.15
	Uzbekistan	0.50	1.00	0.89	0.19
	Central Asia	**0.50**	**1.00**	**0.90**	**0.18**

Note: The unit for the R factor is MJ mm ha^{-1} h^{-1} year^{-1}, and the unit for the K factor is t h MJ^{-1} mm^{-1}. Other factors are dimensionless

erosivity for the baseline period ranges from 41 MJ mm ha^{-1} h^{-1} year^{-1} to 4510 MJ mm ha^{-1} h^{-1} year^{-1}, in the west and southeast, respectively (Fig. 4.1a).

Figure 4.1b and Table 4.1 demonstrate that the erodibility factor in the study area varies from 0.00 to 0.050 t h MJ^{-1} mm^{-1}. The mean erodibility was 0.020 t h MJ^{-1} mm^{-1} with a standard deviation of 0.00 t h MJ^{-1} mm^{-1}. According to the United States Department of Agriculture (USDA), the values range from 0.02 t h MJ^{-1} mm^{-1} for the least erodible soils to 0.64 t h MJ^{-1} mm^{-1} for the most erodible ones.[1]

---

[1] https://www.nrcs.usda.gov/wps/PA_NRCSConsumption/download?cid=stelprdb1262856&ext=pdf

Our results are comparable with those of Kulikov et al. (2020), which indicate that soil erodibility in the study site was 0.0374 t h MJ^{-1} mm^{-1} in mountain rangelands of south–western Kyrgyzstan. This indicates that organic matter plays a significant role in reducing the erodibility of soil (Amanambu et al. 2019; Gupta and Kumar 2017).

A steeply inclined section has a high coefficient of the gradient (Fig. 4.1c). The LS factor values vary from 0.0 to 525.8, with a mean of 4.2 and a standard deviation of 18.8 (Table 4.1). The length and degree of slope inclination are the determining factors in the erosion process. The high LS values of the terrain indicate higher susceptibility to soil erosion.

The C factor was calculated using NDVI to obtain better information about the density of the vegetation cover. It represents better spatial variability of the C factor. The C values have a tendency to grow from the northeast to the southwest of Central Asia. The cover management (C) factor values of the study area range from 0.005 to 0.058. The mean C factor was 0.040, with a standard deviation of 0.008 (Table 4.1). Many researchers have used NDVI to compute the cover management factor. The spatial distribution of the C factor, which is assigned based on the NDVI, is presented in Fig. 4.1d. The P factor values range from 0.5 to 1.0, with a mean of 0.9 and a standard deviation of 0.18 in Kyrgyzstan (Table 4.1).

4.2 Current Rate of Soil Erosion in Central Asia

The results of this work indicate that the mountainous regions of Kyrgyzstan are highly vulnerable to soil erosion. The complex topography and vulnerability to severe semi-arid weather conditions, combined with often poorly monitored pasture practices, make this environment particularly fragile. The average rate of annual soil erosion for the baseline period assessed for the study area varies from <0.01 to >20 t h^{-1} y^{-1}. The mean soil erosion in the study area is 2.62 t h^{-1} y^{-1}. The annual soil loss in Central Asia is 1013.16×10^6 t year^{-1}. Based on the low and high values of the results, the estimated average annual soil erosion in the study area was grouped into five classes (Tables 4.2 and 4.3). The spatial distribution of each class is presented in Fig. 4.2.

4.3 Rainfall Erosivity Analysis Under Baseline and Projected Climate

The WorldClim data (baseline) and observation precipitation were statistically compared. A correlation coefficient of about 0.91 was observed between the baseline and the observed average monthly precipitation. The average annual rainfall erosivity for the observation data ranges from 71.7 to 2390.3 MJ mm ha^{-1} h^{-1} year^{-1} with a mean of 497.8 MJ mm ha^{-1} h^{-1} year^{-1} and standard deviation of 359 MJ mm ha^{-1}

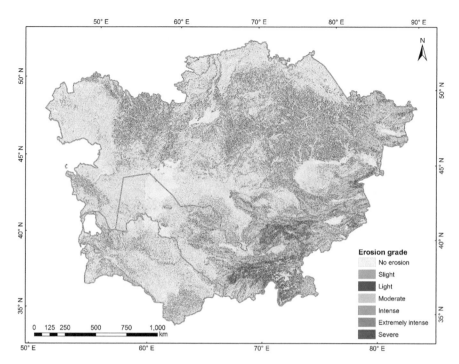

Fig. 4.2 Map of soil erosion grade in Central Asia

Table 4.2 Conversion from erosion rate to erosion grade, with corresponding areas of each erosion grade and its proportion in Central Asia

Soil erosion rate	Area km^2	% share	Mean soil loss (t^{-1} ha^{-1} year^{-1})
Very slight	2494420.75	64.63	0.0002
Slight	550033.88	14.25	0.0440
Light	496120.50	12.86	0.3275
Moderate	132408.75	3.43	2.3180
Intense	43963.06	1.14	7.1744
Very intense	40344.88	1.05	14.3648
Severe	102042.94	2.64	85.6756

The data source for this classification is provided by Teng et al. (2019)

h^{-1} year^{-1}. Conversely, the baseline data show the range of rainfall erosivity to be 95–1838.9 MJ mm ha^{-1} h^{-1} year^{-1} with a mean and standard deviation of 476.8 and 267.1 MJ mm ha^{-1} h^{-1} year^{-1}, respectively. In comparison, the baseline and observed rainfall erosivity produced r, RMSE, and NSE values of 0.90, 156.7 MJ mm ha^{-1} h^{-1} year^{-1}, and 0.60, respectively. This represents a good model performance.

Figure 4.3a presents the relationship between rainfall erosivity with precipitation Moreover, rainfall erosivity has the best correlation with altitude ($r = 0.413$, $r^2 = 0.171$, P < 0.0001), indicating a significant relationship with the spatial variation of

Table 4.3 Estimates of the current annual potential soil loss in Central Asia by countries

Countries	Mean erosion (t^{-1} ha^{-1} $year^{-1}$)	Soil loss ($\times 10^6$ t $year^{-1}$)	Proportion of soil loss of each region (%)	Area (km^2)	% of area
Kazakhstan	0.55	147.47	14.56	2669209.90	69.16
Kyrgyzstan	16.55	292.06	28.83	176512.94	4.57
Tajikistan	39.45	459.38	45.34	116448.06	3.02
Turkmenistan	0.24	11.29	1.11	469008.06	12.15
Uzbekistan	2.40	102.96	10.16	428141.06	11.09
Central Asia	**2.62**	**1013.16**	**100.00**	**3859320.10**	**100.00**

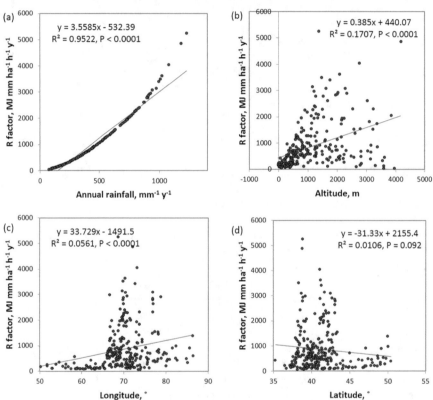

Fig. 4.3 The relationship between rainfall erosivity and annual rainfall (**a**), altitude (**b**), longitude (**c**), and latitude (**d**) in Central Asia

rainfall erosivity in Central Asia. To analyze the influence of the geographic location, a linear relationship was established between the mean annual precipitation erosion and longitude and latitude. As can be seen from Fig. 4.3c, the correlation coefficient between rainfall erosivity and longitude is $r = 0.237$ ($r^2 = 0.056$, P < 0.001), which confirms a significant relationship. From Fig. 4.3d, it can be shown that the coefficient between the rainfall erosivity value and latitude is $r^2 = 0.0106$ (P = 0.092), suggesting an insignificant relationship. Accordingly, the rainfall erosivity increases with the longitude but decreases with the latitude.

The highest annual R factor was observed in the southeastern part of Central Asia, with moderate values in the northern regions, but decreased westwards, where the lowest values were recorded. Conversely, the values gradually increased towards Tajikistan but reduced in the western parts occupied by Turkmenistan. The spatial distribution of the R factor always varied with regard to the annual precipitation in Central Asia. The estimated average annual rainfall erosivity for the baseline period ranges from 41 MJ mm ha^{-1} h^{-1} year^{-1} to 4510 MJ mm ha^{-1} h^{-1} year^{-1} in the west and southeast, respectively (Fig. 4.4).

MIROC5-2.6 and MIROC5-8.5 show higher rainfall erosivity, perhaps due to the strongly projected spatial difference in rainfall in these scenarios. In all the GCMs and baseline precipitation, the R factor in Tajikistan, Kyrgyzstan, east Uzbekistan, and East Kazakhstan is higher but lower in Turkmenistan, northwest Uzbekistan, and southwest and central Kazakhstan (Fig. 4.5). Also, Fig. 4.6 presents the relative difference between the four projected scenarios and the baseline.

Table 4.1 shows the effects of rainfall on historical and projected rainfall erosivity and erosivity density in Central Asia. GCM ensembles express that rainfall erosivity increases significantly from the baseline in all ensembles, except BCCCSM1.1-8.5 in 2070s and IPSLCM5ALR in both RCPs (2030 and 2070). The average value of all scenarios shows that rainfall increased in the 2030s to 262 mm and in the 2070s to 268 mm from the baseline (254 mm). Nevertheless, MIROC5 (the 2030s and 2070s, both RCPs) predicted a higher increase in precipitation than other models with similar scenarios and periods (Table 4.4).

Precipitation, erosivity, and density differ accordingly, given that the GCMs exhibited consistent variations.

The average precipitation and rainfall erosivity demonstrate a steady increase in all the GCMs in combination with the baseline precipitation output. However,

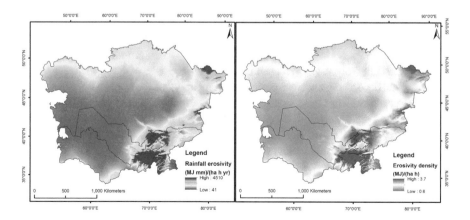

Fig. 4.4 Baseline rainfall erosivity and erosivity density

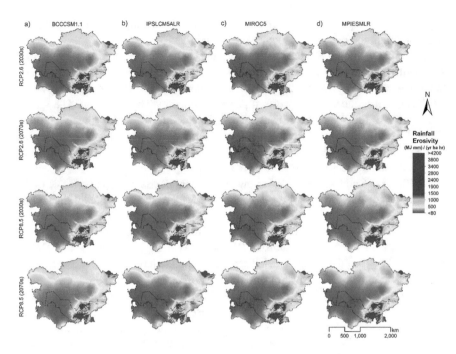

Fig. 4.5 Rainfall erosivity projections for the 2030s and 2070s according to RCP 2.6 and RCP 8.5 scenarios driven by the BCCCSM1.1, IPSLCM5ALR, MIROC-5, and MPIESMLR GCM models

IPSLCM5ALR shows a possible reduction in average precipitation and erosivity in both periods.

4.4   Annual Erosivity Density

The separately projected annual erosivity results (Fig. 4.7) are divided by a corresponding average rainfall data subsequent in the average erosivity density ratio. Density values of erosivity above 1 suggest that a certain amount of precipitation may lead to relatively higher rainfall erosivity (Panagos et al. 2016a). The annual erosivity density for the baseline period has a mean value of 1.38 MJ ha^{-1} h^{-1}, with variability ranging from 0.62 to 3.69 MJ ha^{-1} h^{-1}. The projected variability of erosivity density is also very high because MIROC5-8.5 has the highest mean erosivity density of 1.47 and 1.51 MJ ha^{-1} h^{-1} in the 2030s and 2070s, respectively, followed by MPIESMLR-2.6 with a mean erosivity density of 1.46 MJ ha^{-1} h^{-1} in the 2070s and BCCCSM1.1-2.6 and BCCCSM1.1-8.5 with mean erosivity densities of 1.45 and 1.43 in the 2070s. However, IPSLCM5ALR-2.6 and IPSLCM5ALR-8.5 (both periods) have the lowest mean erosivity density, with an average of 1.35 MJ ha^{-1} h^{-1} (Figs. 4.7 and 4.8).

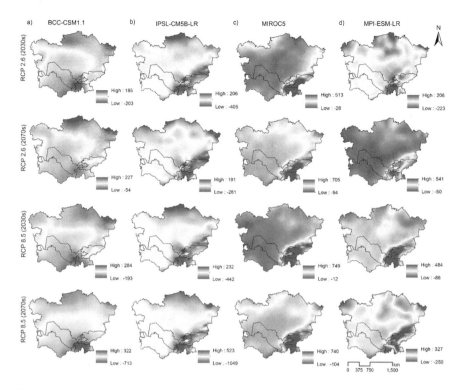

Fig. 4.6 The absolute difference in rainfall erosivity (MJ mm ha^{-1} h^{-1} year^{-1}) between the 2030 and 2070 projections and baseline data

4.5   Projected Future Soil Erosion Potential in Central Asia

According to the results of the soil erosion map (Fig. 4.9), most of the study area had a low risk of erosion in the baseline period. By substituting rainfall erosivity and taking into account all other constant factors in the RUSLE model, soil erosion was predicted and calculated for the different GCMs (BCCCSM1.1, IPSLCM5ALR, MIROC-5, and MPIESMLR) as well as for RCP 2.6 and RCP 8.5). The results of the above modeling are presented in Fig. 4.9. In Central Asia, increasing soil erosion is reflected by the variations in annual erosivity, which is evident in all scenarios and shows a significant growth trend. However, in the 2030s, BCCCSM1.1-2.6 and BCCCSM1.1-8.5 did not show a significant change. In the 2030s, MIROC5-2.6 and MIROC5-8.5 display the highest percentage of growth in soil loss of about 16.1% and 15.6%, respectively. Despite the high R factor, BCCCSM1.1-8.5 (2070s) shows a decrease in soil erosion (−13.3%). During this climatic period, the highest rainfall erosivity is observed at MIROC5-8.5, where the annual soil loss was about 3.33 t ha^{-1} year^{-1}, which increased from the baseline soil loss by 20.82%.

In general, there is a growth in the average annual soil loss for most model ensembles in both climatic periods (2030s and 2070s). Figures 4.9, 4.10, and 4.11 show the

Table 4.4 Changes in average rainfall erosivity, erosion, and erosivity density under climate change across Central Asia

Climate models	Precipitation	Rainfall erosivity (MJ mm ha^{-1} hr^{-1} year^{-1})	Change (%)	Erosivity density	Change (%)
Baseline (1961–1990)	253.57	402.07	0	1.38	0
2020–2039					
BCCCSM1.1-2.6	263.5	430.01	6.95	1.41	2.2
BCCCSM1.1-8.5	267.12	437.07	8.7	1.42	2.9
IPSLCM5ALR-2.6	247.31	386.65	-3.84	1.36	-1.4
IPSLCM5ALR-8.5	246.48	386.37	-3.9	1.35	-2.2
MIROC5-2.6	266.4	439.64	9.34	1.42	2.9
MIROC5-8.5	283.19	481.98	19.87	1.47	6.5
MPIESMLR-2.6	254.36	404.09	0.5	1.38	0.0
MPIESMLR-8.5	263.94	430.14	6.98	1.41	2.2
Average	261.54	424.49	5.58	1.4	1.6
2060–2079					
BCCCSM1.1-2.6	273.95	450.35	12.01	1.45	5.1
BCC-CSM1.1-8.5	268.61	437.77	8.88	1.43	3.6
IPSLCM5ALR-2.6	248.82	391.22	-2.7	1.36	-1.4
IPSLCM5ALR-8.5	243.9	381.36	-5.15	1.34	-2.9
MIROC5-2.6	270.33	449.88	11.89	1.43	3.6
MIROC5-8.5	294.11	508.85	26.56	1.51	9.4
MPIESMLR-2.6	278.9	469.3	16.72	1.46	5.8
MPIESMLR-8.5	267.4	435.84	8.4	1.42	2.9
Average	268.25	440.57	9.58	1.43	3.3

spatial change in potential annual soil erosion and absolute differences in Central Asia, which is estimated by the RUSLE model in various projected climatic scenarios. These increases in annual soil loss may be due to a blend of several factors. Of note, however, is that changes in rainfall erosivity often occur because of changes in annual rainfall. The results of the average changes in soil loss (Table 4.5) in the historical and projected climate change predictions from four various climate models for the two different periods are 2.72 t ha^{-1} year^{-1} (2030s) and 2.71 t ha^{-1} year^{-1} (2070s).

The annual soil erosion shows a corresponding increase of about 20.82% from the baseline climate for the same time slice and climate scenario. MPIELSMLR-8.5 also shows a decrease in soil erosion (−1.47%), even though the erosivity is reported to be high. For the 2020 to 2039 climatic period, MIROC5-2.6 and MIROC5-8.5 show the highest positive percentage increases, in soil erosion, of about 12.02% and 20.23% respectively. As rainfall erosivity increases from the baseline erosion, the percentage of changes in annual erosion show increases.

The dynamic influence of climate change on soil erosion is another essential factor that is uncertain; however, it may depend on the interacting effects of the associated factors. Yet, in the climate change-based soil erosivity predictions, it can be seen that the average soil loss, resulting from the RUSLE model within the ensem-

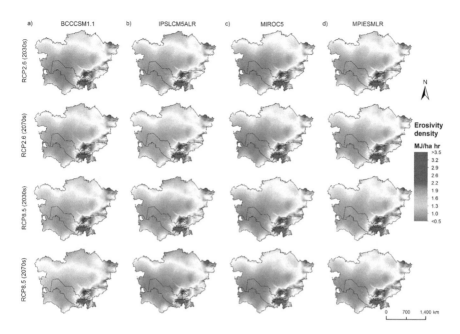

Fig. 4.7 Erosivity density for the different scenarios for the 2030s and 2070s according to RCP 2.6 and RCP 8.5 scenarios driven by the BCCCSM1.1, IPSLCM5ALR, MIROC-5, and MPIESMLR GCM models

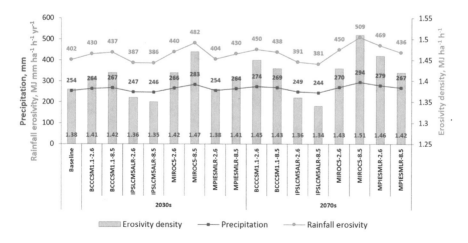

Fig. 4.8 Annual precipitation, R factor, and erosivity density for different scenarios and periods

ble combinations, show some disparities. However, future soil erosion rates are expected to increase due to increased precipitation and rainfall erosivity. This has been confirmed in other prediction studies (e.g., Amanambu et al. (2019); Gafforov et al. (2020); Gupta and Kumar (2017); Litschert et al. (2014); Plangoen et al. (2013)) which demonstrate that soil erosion rates will be significantly impacted by

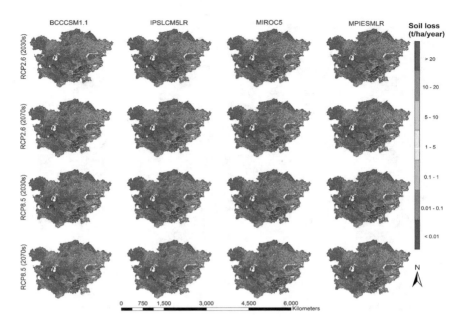

Fig. 4.9 Soil erosion for different scenarios for the 2030s and 2070s according to the RCP 2.6 and RCP 8.5 scenarios driven by the BCCCSM1.1, IPSLCM5ALR, MIROC-5, and MPIESMLR GCM models

an increase in precipitation and intensity. The highest percentage of soil loss occurs in the medium and high regions of Tajikistan, Kyrgyzstan, Eastern Uzbekistan, and East Kazakhstan. This suggests that there will be more occasions for soil losses at medium and high altitudes than has ever been experienced in the past. Consequently, high soil erosion may lead to high sedimentation in rivers, lakes, and reservoirs, and these are critical for flooding and water pollution (Yang et al. 2003) (Fig. 4.12).

4.6 Rainfall Erosivity and Soil Erosion at the National Level

In the baseline period, Kyrgyzstan has an estimated average rainfall erosivity of 869.7 MJ mm ha^{-1} h^{-1} year^{-1}. The MIROC5 and MPIESMLR scenarios (RCP 2.6 and RCP 8.5, respectively) project an increase in the mean rainfall erosivity ranging from 27.9% to 50.1% and from 0.9% to 27%. The BCCCSM1.1 scenarios also projected a mean increase of 6.3% for all periods and a decrease (-7.8%) for BCCCSM1.1-8.5 in the 2070s.

In Kazakhstan, we calculated the mean values of rainfall erosivity of 374.3 MJ mm ha^{-1} h^{-1} year^{-1} during the baseline period. From MIROC5, BCCCSM1.1, and MPIESMLR (both scenarios), we note an increase in the mean rainfall erosivity of 5.3% to 27.3%, 11.6% to 23.2%, and 0.6% to 15.7%, respectively, in this country with a slight increase in the northern part and a significant increase in the eastern

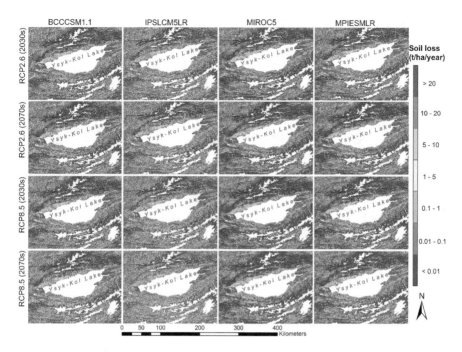

Fig. 4.10 The areas around Ysyk-Kol Lake (Kyrgyzstan) with severe soil erosion. Soil erosion for different scenarios for the 2030s and 2070s according to the RCP 2.6 and RCP 8.5 scenarios driven by the BCCCSM1.1, IPSLCM5ALR, MIROC-5, and MPIESMLR GCM models

part of this country. We also found results using IPSLCM5ALR for both scenarios, with decreases ranging from −0.3% to 1.6%, except RCP 8.5 (2070s), which has an increase of 1.5%. The most significant increase in rainfall erosivity is expected for the 2060–2079 period using the RCP 8.5 scenario (Fig. 4.13 and Table 4.6).

The average rainfall erosivity in Turkmenistan for the baseline period was 188.4 MJ mm ha^{-1} h^{-1} year^{-1}. This country has the lowest rainfall erosivity among all Central Asian countries. The ensemble scenarios of IPSLCM5ALR and BCCCSM1.1 (RCP 2.6 and RCP 8.5) predict decreases from −10.1% to −19% and −7.1% to −16%, respectively. However, BCCCSM1.1-2.6 (2070s) predicts about a 10.5% increase in rainfall erosivity. The MPIESMLR and MIROC5 results indicate changes from 2.3% to 14.3% and 3.7% to 19.6%, whereas MIROC5-2.6 (2070s) decreased (−4.7%) during the two-time slices for the two emission scenarios.

For Tajikistan, an average rainfall erosivity of 1447.7 MJ mm ha^{-1} h^{-1} year^{-1} in the baseline period was revealed. The increase was observed in the MIROC5 and MPIESMLR scenarios, thereby indicating the highest rainfall erosivity in the study area. However, there is also a decrease in the average rainfall erosivity in this country for IPSLCM5ALR (both scenarios and both periods), BCCCSM1.1-8.5 (both periods), and MPIESMLR-2.6 (2030s) compared with the baseline. For both scenarios, MIROC5 projected an increase in erosivity from 34.3% to 56.3%, whereas IPSLCM5ALR projected a decrease from −8.2% to −26.2%.

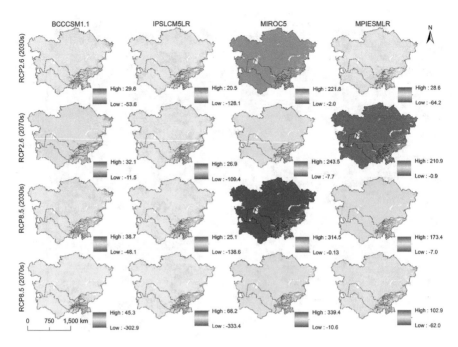

Fig. 4.11 The absolute difference in soil erosion (t ha^{-1} year^{-1}) between the 2030 and 2070 projections and baseline data

Table 4.5 Changes in average erosion under climate change averaged across Central Asia

Periods	RCPs	Models	Mean soil loss (t^{-1} ha^{-1} year^{-1})	Differences	Change (%)
		Baseline	2.62	0.00	0.00
2030s	RCP 2.6	BCCCSM1.1	2.62	0.00	0.00
		IPSLCM5LR	2.33	−0.29	−8.50
		MIROC5	3.03	0.41	12.02
		MPIESMLR	2.57	−0.05	−1.47
	RCP 8.5	BCCCSM1.1	2.62	0.00	0.00
		IPSLCM5LR	2.28	−0.34	−9.97
		MIROC5	3.31	0.69	20.23
		MPIESMLR	2.99	0.37	10.85
2070s	RCP 2.6	BCCCSM1.1	2.70	0.08	2.35
		IPSLCM5LR	2.44	−0.18	−5.28
		MIROC5	3.06	0.44	12.90
		MPIESMLR	3.09	0.47	13.78
	RCP 8.5	BCCCSM1.1	2.30	−0.32	−9.38
		IPSLCM5LR	1.95	−0.67	−19.65
		MIROC5	3.33	0.71	20.82
		MPIESMLR	2.80	0.18	5.28

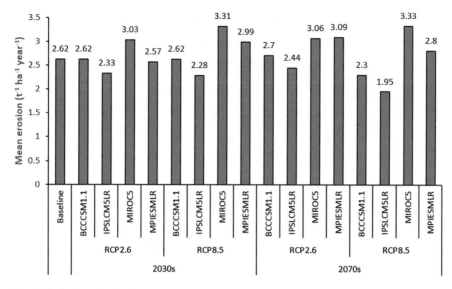

Fig. 4.12 Soil erosion in Central Asia for different scenarios and periods

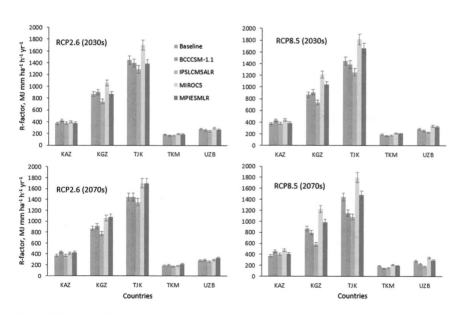

Fig. 4.13 Average R factor values projected by the GCMs under two RCPs for Central Asian countries over different time slices

Table 4.6 Rainfall erosivity in Central Asia by countries. Mean baseline and estimated (MJ mm ha^{-1} h^{-1} year^{-1}) by BCCCSM1.1, IPSLCM5ALR, MIROC5, and MPIESMLR with RCP 2.6 and RCP 8.5 emission scenario models. Projected change with regard to baseline (%)

		Kyrgyzstan	Kazakhstan	Tajikistan	Turkmenistan	Uzbekistan
Baseline (1961–1990)		869.7	374.3	1447.7	188.4	282.1
RCP 2.6 (2030s)	BCCCSM-1.1	903.8	420.3	1395.9	167.7	258
	Change, %	6.3	11.6	0.4	−10.1	−8.1
	IPSLCM5ALR	744.1	377.5	1290.8	164.7	240.5
	Change, %	−17.2	−0.3	−11	−11.2	−11.9
	MIROC5	1057	400.6	1697.1	196.9	295.6
	Change, %	27.9	5.3	36.9	3.7	2.1
	MPIESMLR	870.1	380.8	1388.3	191.2	272.5
	Change, %	0.9	0.6	−2.2	2.3	−2.7
RCP 2.6 (2070s)	BCCCSM-1.1	910	436.2	1450.2	200.9	291.9
	Change, %	6.3	17.7	3.8	10.5	9
	IPSLCM5ALR	769	375.1	1350.5	172.9	263.4
	Change, %	−14.5	−1.5	−8.2	−7.1	−7.7
	MIROC5	1062.6	416.8	1699.9	179.9	297
	Change, %	28.8	9.8	34.3	−4.7	2.4
	MPIESMLR	1081.9	432.5	1702.5	216.9	329
	Change, %	27	15.7	24.5	14.3	15.3
RCP 8.5 (2030s)	BCCCSM-1.1	909.6	430.9	1386.2	165	256.6
	Change, %	6.3	14.9	−5.7	−11.1	−10
	IPSLCM5ALR	736.4	380.2	1254.8	171	227.6
	Change, %	−17.4	−1.6	−15.4	−7.3	−17.7
	MIROC5	1216.8	435.9	1818.4	212	336.9
	Change, %	49.7	17	51.3	13.5	19.4
	MPIESMLR	1041.7	385.2	1664.4	202.8	320
	Change, %	22	4.6	20.8	8.8	14.5
RCP 8.5 (2070s)	BCCCSM-1.1	794	456.8	1150.6	146.9	228.4
	Change, %	−7.8	23.2	−26.3	−19	−17.7
	IPSLCM5ALR	581.3	400	1082	154.2	182.8
	Change, %	−35.5	1.5	−26.2	−16.7	−28.2
	MIROC5	1224.1	473.1	1800.4	215.3	340
	Change, %	50.1	27.3	56.3	19.6	24.1
	MPIESMLR	988	409.6	1484.7	195.4	294.5
	Change, %	15.7	11.1	9	7.6	7

Uzbekistan had an average baseline rainfall erosivity of 282.1 MJ mm ha^{-1} h^{-1} year^{-1}. The MIROC5 scenarios (RCP 2.6 and RCP 8.5) projected an average increase ranging from 2.1% to 24.1%, whereas MPIESMLR projected a mean increase ranging from 7% to 15.3%, except for a decrease of −2.7% for the RCP 2.6 (the 2030s). IPSLCM5ALR projected a decrease in the mean annual rainfall erosivity from −7.7% to 28.2%. In general, all scenarios projected a substantial increase and decrease in rainfall erosivity over Uzbekistan (Table 4.2). Our results (<280 MJ mm ha^{-1} h^{-1} year^{-1}) are comparable with those of Panagos et al. (2017b), who reported low-average erosivity values (<250 MJ mm ha^{-1} h^{-1} year^{-1}) in Kazakhstan, Turkmenistan, and Uzbekistan.

In Table 4.7, the average soil erosion is presented per country for two periods and RCPs. Tajikistan indicated the highest soil erosion, which is a consequence of the rainfall amount in the Pamir Mountains. The lowest values were observed in Turkmenistan, Uzbekistan, and Kazakhstan.

Table 4.7 Soil erosion in Central Asia by country. Mean baseline and estimated (t ha^{-1} year^{-1}) by BCCCSM1.1, IPSLCM5ALR, MIROC5, and MPIESMLR with RCP 2.6 and RCP 8.5 emission scenario models. Projected change with regard to baseline (%)

		Kazakhstan	Kyrgyzstan	Tajikistan	Turkmenistan	Uzbekistan
Baseline (1961–1990)		0.55	16.55	39.45	0.24	2.40
RCP 2.6 (2020–2039)	BCCCSM-1.1	0.59	16.98	38.16	0.22	2.25
	Change, %	7.60	2.62	−3.26	−9.14	−6.36
	IPSLCM5ALR	0.51	14.36	35.35	0.21	2.06
	Change, %	−6.88	−13.21	−10.39	−14.65	−14.35
	MIROC5	0.60	19.72	46.41	0.25	2.52
	Change, %	9.29	19.18	17.65	5.36	4.58
	MPIESMLR	0.56	16.40	38.07	0.24	2.32
	Change, %	0.73	−0.91	−3.49	−0.19	−3.36
RCP 2.6 (2060–2079)	BCCCSM-1.1	0.61	17.17	39.54	0.25	2.40
	Change, %	10.44	3.74	0.23	4.05	−0.14
	IPSLCM5ALR	0.54	14.80	36.81	0.22	2.29
	Change, %	−2.89	−10.57	−6.69	−9.34	−4.93
	MIROC5	0.62	19.71	46.93	0.23	2.53
	Change, %	12.85	19.12	18.98	−4.30	5.24
	MPIESMLR	0.62	19.71	46.93	0.23	2.53
	Change, %	12.85	19.12	18.98	−4.30	5.24
RCP 8.5 (2020–2039)	BCCCSM-1.1	0.61	17.13	37.80	0.22	2.28
	Change, %	9.90	3.52	−4.18	−9.91	−5.32
	IPSLCM5ALR	0.50	14.16	34.42	0.21	2.01
	Change, %	−9.11	−14.44	−12.74	−11.92	−16.58
	MIROC5	0.65	22.34	49.67	0.27	2.74
	Change, %	18.06	35.05	25.91	12.25	13.77
	MPIESMLR	0.58	19.57	45.37	0.26	2.64
	Change, %	5.43	18.28	15.02	7.67	9.77
RCP 8.5 (2060–2079)	BCCCSM-1.1	0.60	14.92	31.32	0.19	2.05
	Change, %	9.42	−9.84	−20.60	−19.22	−14.95
	IPSLCM5ALR	0.50	11.20	29.76	0.19	1.52
	Change, %	−9.87	−32.34	−24.56	−21.63	−36.88
	MIROC5	0.69	22.45	49.64	0.26	2.67
	Change, %	24.36	35.68	25.84	8.08	11.19
	MPIESMLR	0.62	18.49	40.77	0.24	2.43
	Change, %	11.51	11.73	3.34	−0.12	1.00

A remote area that is sensitive to climate variability and soil erosion in the whole Central Asia has been rarely reported quantitatively, and few research studies have provided a prediction of the future soil erosion risk in Central Asia. In this study, soil erosion in Central Asia is based on the RUSLE model. The influence of climate change on rainfall erosivity is expressed by variations in total precipitation, as presented in the Results section. High altitude strongly influences the rate of erosivity in the study area. The Tien Shan, Pamir-Alay, and Pamir mountains experience more torrential rainfall compared with the surrounding low-lying deserts. The erosion pattern detected in this paper is attributed to the drier climate of the southwest and or western lowlands, and the southeast and or eastern highlands, which is wet throughout the year.

The changes in precipitation mainly depend on the changes in the water content in the atmosphere, which is transferred from the oceans to the earth through a large-scale atmospheric circulation (Luo et al. 2019). The atmospheric circulation over Central Asia is characterized by the predominance of the west–east transfer of air masses when the main moisture that gives precipitation comes from the North Atlantic Ocean (Alamanov et al. 2006; Chen et al. 2018; Hu et al. 2018; Hu et al. 2017). Most of the low-latitude region (40° N) is characterized by low-pressure anomalies (Hu et al. 2018). As the air masses move from the Atlantic Ocean, they lose moisture to become dry air mass as they approach the territory, causing little or no precipitation in summer (Alamanov et al. 2006).

El Niño–Southern Oscillation (ENSO) has affected precipitation changes over the arid regions of Central Asia by the southwestward flow of water vapor coming from the Arabian Sea and tropical Africa (Chen et al. 2018; Hu et al. 2017; Mariotti 2007). ENSO-induced precipitation is related to large-scale atmospheric circulation changes caused by sea surface temperature (SST) (Dai and Wigley 2000). Previous studies have shown that changes in SST have a significant effect on the transport of water vapor from the oceans to land (Liu and Duan 2017; Luo et al. 2019).

The main feature in the distribution of precipitation in Central Asia is their small annual amount in the lower part of the territory, which results in the formation of vast deserts. At the same time, on the shores of the Caspian Sea, and especially Balkhash Lake, precipitation is usually low. Only in the mountains, on the outer windward slopes, where the air masses experience a forced rise and, as a result, cool down, reaching a state of saturation, does the orographic increase in precipitation occur three to five times or more compared with the surrounding deserts (Alamanov et al. 2006). This fact explains the spatial distribution of precipitation in Central Asia and, in turn, may clarify why some parts have higher erosion than the other parts. In addition, urbanization and industrialization in the densely populated Fergana Valley (southeast) directly contribute to climate change in the form of regional warming. This results in precipitation variability that consequently influences erosion.

The mountain system, located in the adjacent areas of Central Asia, has a significant influence on the climate of the region. It includes not only barriers protecting Central Asia from the south (from the penetration of monsoons in South Asia). They form a barrier of over 3000-km long and have a distorting effect on the height of frontal zones, which significantly affect the development of cyclonic activity in Central Asia. The mountain barrier significantly reduces the amount of precipitation brought by the West, in particular by southwestern (Mediterranean) cyclones to Central Asia (Issanova and Abuduwaili 2017).

As stated by Amanambu et al. (2019), erosion is influenced by the changes in precipitation patterns and quantities due to climate change. Studies in the Eurasian continent predicted a significant increase or decrease in erosivity for the future climate. For example, Panagos et al. (2017a) found that 81% of the territory in Europe is projected to have an increase in rainfall erosivity, and it is projected that by 2050, there will be a 19% decrease in rainfall erosivity (HadGEM2, RCP 4.5 scenario). Likewise, however, our study predicts some spatial variability in erosivity for Central Asia concerning the anthropogenic influence on the amount of precipitation based on different GCMs.

Besides, the powerful influence of climate change on soil erosion is another essential factor that is uncertain; however, it may depend on the interacting effects of the associated factors. In the climate change-based soil erosivity predictions, it can be observed that the average soil loss, resulting from the RUSLE model within the ensemble combinations, yet show some disparities. However, future soil erosion rates are expected to increase due to increased precipitation and rainfall erosivity. This has been confirmed in other prediction studies (Amanambu et al. 2019; Gupta and Kumar 2017; Litschert et al. 2014; Plangoen et al. 2013), which demonstrate that an increase in precipitation and intensity will significantly impact soil erosion rates.

Small percent variations are usually expected in developed areas, which generally have lower hills of slope and height. The variability of our results shows the disagreements in scenarios, periods, and climate models but may also show persistent uncertainty in our models. The WorldClim precipitation data were used together with equations based on the annual precipitation to calculate precipitation erosion rates. The number of weather stations in the Central Asian region is limited, and the resolution of spatial data on rainfall is low. Besides, this approach does not cover the distribution of heavy rains, which are known to have a significant effect on soil erosion. The product of the total energy of the storm and maximum intensity of the storm mainly determine the potential of a 30-min rainfall to cause erosion. There is no detailed rainfall data at sub-hourly intervals for Central Asia.

The preservation of fertile soils by agricultural lands, pastures, and forests is the primary condition for the sustainable development of humanity. The possible increase in rainfall erosivity in Tajikistan and Kyrgyzstan may affect a significant part of the agricultural production in Central Asia due to increased soil loss and reduced soil fertility and water availability. Conversely, a reduction in the rainfall erosivity over Turkmenistan and western Uzbekistan can strengthen the trend of agricultural development in these areas. However, climate change can significantly affect the land cover, which can balance or reinforce some erosion trends. To predict future soil erosion trends, these feedbacks between precipitation and land cover should be evaluated.

A prediction of future soil erosion in Central Asia, using four GCMs, shows that southwestern Central Asia appears to be a region that is more sensitive to climate change in both periods, especially if the conditions of the RCP 8.5 scenario arise, which correspond to the path with the greatest climatic variability and the highest greenhouse gas emissions. The trend towards an increase in total soil runoff in this area under the RCP 2.6 and RCP 8.5 scenarios indicates a future risk of erosion in the 2030s and 2070s. The occurrence of increased soil erosion can affect local eco-systems in Central Asia and, consequently, cause hydrological changes in the rivers originating in the mountainous areas in Tajikistan and Kyrgyzstan, such as the Amu Darya, Syr Darya, and Chu and Talas rivers.

We used two emission scenarios for future projections that fall at the lowest and highest ends of all warming scenarios. However, how much warming will happen in Central Asia is still unknown. The results of soil erosion predicted by the four GCMs exhibit different trends in some regions in Central Asia, and this is reflected in the high uncertainty of the projected future soil erosion. We believe that the scenarios and projection models we used would provide a useful threshold for soil erosion in the 2030s and 2070s.

Chapter 5
Spatiotemporal Variations and Projected Rainfall Erosivity and Erosivity Density in Kazakhstan

Abstract In this chapter, the spatial-temporal variation of rainfall erosivity in Kazakhstan in 1970–2017 was investigated. The results showed that the average annual rainfall erosivity in Kazakhstan over the past 48 years was 464 MJ mm ha^{-1} h^{-1} year^{-1}. No significant time trend was found in annual rainfall erosivity. Some of the results presented here are relevant to the further study of potential soil erosion in Kazakhstan. The East Kazakhstan, North Kazakhstan, Almaty regions were under a more significant threat of rainfall erosivity than other regions. It is important to understand past and future differences in rainfall erosivity and its consequences in arid and semi-arid regions, where the amount of daily precipitation is always limited. GCM scenarios (GISSE2H, HadGEM2-ES, and NorESM1M) were statistically downscaled using the delta method for three periods. This study estimated the long-term variations in annual rainfall erosivity in Kazakhstan using past and future climate data. Based on the baseline climate, the average change in percent rainfall erosivity is 26.9%, 26.4%, and 35.2% in the 2030s, 2050s, and 2070s, respectively. The aggregate average annual precipitation and erosion activity for all climate models for all scenarios shows steady growth compared with the baseline climate.

Keywords Kazakhstan · Rainfall erosivity · Erosivity density · Spatiotemporal change · Climate change · GCMs · RCPs · Trends · Baseline · Future scenarios

5.1 Physical–Geographical Characteristics of Kazakhstan

Kazakhstan is the largest country in Central Asia; its area is 2.7 million km^2, and it has 14 administrative regions (Koshim et al. 2018). Kazakhstan borders with Russia in the north; with China in the east; with Kyrgyzstan, Uzbekistan, and Turkmenistan in the south; and with the Caspian Sea in the west. The Tien Shan and Altai Mountains are located in eastern and southeastern Kazakhstan. The northern, western, and northwestern regions consist of plains, whereas the central part is hilly. The

E. Duulatov et al., *Current and Future Trends of Rainfall Erosivity and Soil Erosion in Central Asia*, SpringerBriefs in Environmental Science, https://doi.org/10.1007/978-3-030-63509-1_5

general landform in Kazakhstan is mountainous terrain in the east and southeast and lowland in the west and northwest (Liang et al. 2017). The territory of Kazakhstan is far from the oceans and is open to winds from the north. In addition, the climate of Kazakhstan is continental, with an uneven distribution of precipitation. The flat areas are usually dry and have annual precipitation from 100 mm in the southwest to 400 mm in the north. In mountainous areas, the rainfall ranges from 400 to 1600 mm (Issanova et al. 2015). The average January temperature is −18 °C in the north and −3 °C in the south. The average temperature in July gradually rises from 19 °C in the north to 28–30 °C in the south (Issanova et al. 2015). The main types of ecosystems in Kazakhstan are meadows, arable land, and shrubs. From north to south, the type of land consistently changes from arable land to pastures and shrubs. Arable land is mainly distributed in the southern and northern border areas. Grasslands are mainly distributed in the central areas in the north, west, and northwest and shrubs in the south and south (Liang et al. 2017) (Fig. 5.1).

5.2   Annual Rainfall Erosivity Analysis and Trend

The average annual rainfall erosivity for each meteorological station was obtained by annual precipitation calculated using Eqs. (3.5) and (3.6). The results show that Kazakhstan's annual rainfall erosivity is 467 MJ mm ha^{-1} h^{-1} year^{-1} on average. The standard deviation reached 92 MJ mm ha^{-1} h^{-1} year^{-1}, indicating that spatial varia-

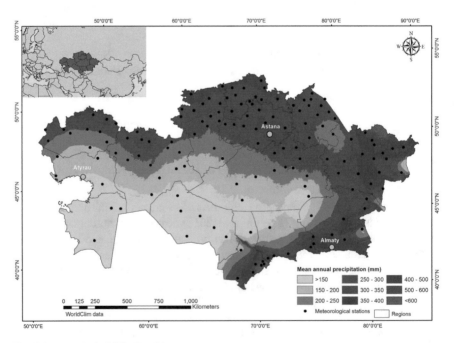

Fig. 5.1 Annual rainfall in Kazakhstan

tion was considerable. Relatively larger values were mainly distributed in the South Kazakhstan, as well as Almaty, Petropavl, Zhambyl, Akmola, and the East Kazakhstan regions. Rainfall erosivity in Kostanay, Pavlodar, West Kazakhstan, Aktobe, and Karagandy is at the middle level, and that of Atyrau, Mangystau, and Kyzyl-Orda is at the lower level. From east to west, rainfall erosivity has declined overall.

Figure 5.2a and 5.2b present the relationship between rainfall erosivity and precipitation and between rainfall erosivity and erosivity density, respectively. Appendix A presents the rainfall, rainfall erosivity, and erosivity density from 150 stations in Kazakhstan.

A linear relationship was established between the mean annual precipitation erosion and longitude/latitude. According to Fig. 5.3a, the correlation coefficient between rainfall erosivity and longitude is $r = r^2$ (0.0734) = 0.271 (P < 0.001), confirming a significant relationship. Figure 5.3b present the coefficient between rainfall erosivity value and latitude as $r = r^2$ (0.0143) = −0.120 (P = 0.144), suggesting an insignificant relationship. Accordingly, rainfall erosivity increases with longitude but decreases with latitude (Table 5.1).

Moreover, rainfall erosivity has the best correlation with altitude ($r = r^2$ (0.3562) = 0.597, P < 0.0001), indicating a significant relationship with the spatial variation of rainfall erosivity in Kazakhstan (Fig. 5.4).

Annual Rainfall Erosivity Trend   Figure 5.6 indicates that the annual rainfall erosivity over Kazakhstan increased in the last 48 years and that 105 stations were in the insignificant increasing trend and 45 stations were in an insignificant decreasing trend. At a significance level of α = 0.05 (95% confidence level), the variation trends can be analyzed and discussed by the correlation coefficients. Rainfall erosivity of most areas of the basin exhibited an insignificant upward trend, indicating that the probability of water and soil erosion caused by rainfall is continuously increasing. Rainfall erosivity was calculated based on the average annual precipitation from 1970 to 2017 (Fig. 5.5). The variations in trends were analyzed using linear regression for the year timescale level. The critical value of the correlation coefficient test

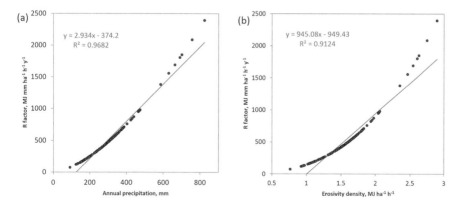

Fig. 5.2 Relation R factor with (**a**) precipitation and (**b**) erosivity density

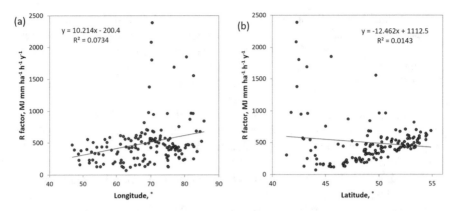

Fig. 5.3 The relationships of rainfall erosivity with longitude (**a**) and latitude (**b**)

Table 5.1 Correlation matrix (Pearson) of rainfall erosivity with erosivity density, precipitation, altitude, longitude, and latitude

Variables	R factor	Erosivity density	Precipitation	Altitude	Latitude	Longitude
R factor	1	0.955	0.984	0.597	-0.120	0.271
Erosivity density		1	0.993	0.542	0.063	0.310
Precipitation			1	0.569	-0.012	0.299
Altitude				1	-0.442	0.524
Latitude					1	-0.046
Longitude						1

Values in bold are different from 0 with a significance level alpha=0.05

Fig. 5.4 The relationship of rainfall erosivity with altitude

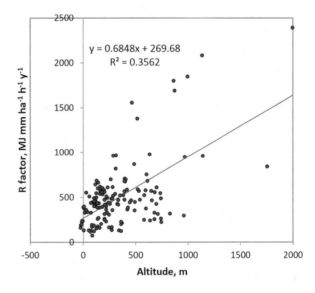

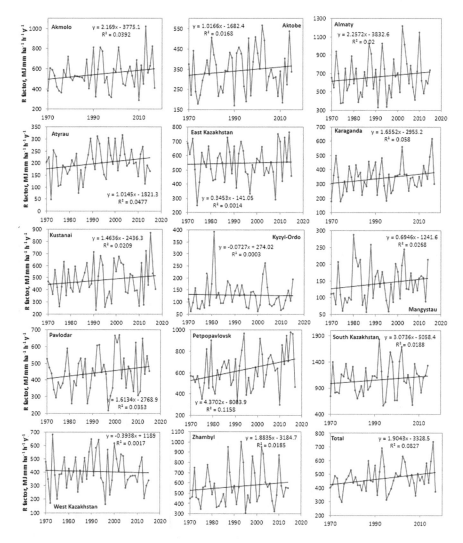

Fig. 5.5 The inter-annual variation of the annual rainfall erosivity in Kazakhstan at the regional level

r is 0.0827 at the significance level. Rainfall erosivity in most regions of Kazakhstan exhibits a tendency to increase, indicating that the possibility of water and soil erosion caused by precipitation is continuously increasing.

Precipitation is one of the main external factors contributing to soil erosion. Therefore, in this section, we quantitatively discuss the relationship between rainfall erosivity and precipitation. Rainfall erosivity for most stations exhibited an upward trend. For further analysis of the correlation between them, a linear relationship was established in Fig. 5.2. The calculation results show that the correlation coefficient between them is $r = r2\ (0.968) = 0.984$ (P < 0.0001), proving that

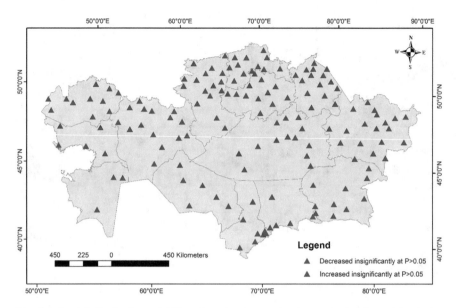

Fig. 5.6 Spatial distributions of 150 stations for the trends of annual rainfall erosivity during 1970–2017 in Kazakhstan

they pass the significance test. It is noteworthy that this data from the outside explains that we can approximately estimate the distribution of annual rainfall erosivity by the distribution of annual rainfall and qualitatively estimate the annual value of rainfall erosivity by the amount of precipitation, which provides a simple method for estimating rainfall erosivity. Soil loss is proportional to rainfall erosivity (R) when other factors are kept constant (Gu et al. 2018).

The annual rainfall erosivity in Kazakhstan exhibited an insignificant temporal trend. At most stations, rainfall erosivity exhibited an upward trend (Fig. 5.6). This implies a future, further deterioration of water and soil loss caused by internal and external (human) factors. Thus, fruitful, productive, and calculated conservation measures are important to protect the water and soil resources in Kazakhstan. Statistical analysis revealed that the annual rainfall erosivity in Kazakhstan during 1970–2017 spatially varied. The above analysis suggests that the soil and water loss situation in Kazakhstan is optimistic (Table 5.2).

5.3   Projected Impacts of Climate Change on Rainfall Erosivity

The simulated (WorldClim) and observed precipitations (1970–2017) were statistically and visually compared to summarize the strength with which the simulated WorldClim data (1960–1990) can reproduce the spatial pattern of the mean and

Table 5.2 Rainfall erosivity in 48 years in Kazakhstan and its regions (MJ mm ha^{-1} h^{-1} year^{-1})

Regions	Years	Minimum	Maximum	Mean	Std. deviation
Akmola	48	287.1	1023.8	548.8	153.3
Aktobe	46	169.8	567.6	343.2	105.3
Almaty	46	330.6	1220.8	664.8	214.5
Atyrau	46	47.8	317.2	200.1	62.3
East Kazakhstan	46	251.6	766.3	547.0	123.3
Karagandy	48	177.5	618.8	344.4	96.2
Kostanay	48	234.4	870.9	481.4	141.9
Kyzyl-Orda	46	60.9	392.4	129.1	60.9
Mangystau	45	57.7	288.1	142.1	55.7
Pavlodar	46	217.8	669.9	445.7	115.2
Petropavl	48	298.5	978.1	628.1	179.8
South Kazakhstan	45	570.5	1831.1	1064.2	294.3
West Kazakhstan	46	164.6	680.0	404.4	128.2
Zhambyl	45	307.8	1001.5	567.1	181.7
Total	48	298.1	737.3	467.7	92.7

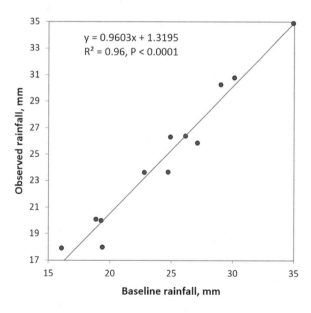

Fig. 5.7 Scatter plot between the observed and baseline average monthly rainfall

y = 0.9603x + 1.3195
R^2 = 0.96, P < 0.0001

annual precipitation (Fig. 5.1). The scatter plot (Fig. 5.7) demonstrates the quality of the simulated dataset compared with the observed dataset for all 150 stations, with an r^2 of about 0.96. A correlation coefficient of about 0.98 between the simulated and observed monthly precipitations was observed. This indicates that the simulated baseline precipitation reproduces currently observed precipitation with considerable accuracy. Table 5.3, Figs. 5.8, and 5.9 show rainfall erosivity (R) for the observation data for the 150 meteorological stations and the R factor for the baseline data in the stations.

Table 5.3 Rainfall erosivity (MJ mm ha^{-1} h^{-1} year^{-1}) for the baseline and observation in Kazakhstan, from 150 meteorological stations

Stations	Observed	Baseline	Stations	Observed	Baseline	Stations	Observed	Baseline
Petropavl	701.3	526.7	Pavlodar	454.7	415.8	Chapaevo	399	377.9
Ruzaevka	609.2	492.9	Sharbakty	432.8	413.4	Karatyuba	336.6	321.6
Saumalkol	819.2	545.3	Shalday	385.1	384.9	Uralsk	554	529.4
Sergeevka	663.5	534.7	Koktobe	387.9	355	Aksai	509.1	526.7
Blagoveschenka	600.9	580.4	Besoba	314.2	415.8	Chingirlau	435.9	455
Yavlenka	641.4	556	Korneevka	576.8	475.1	Zhambeity	358.3	377.9
Timiryazevo	565.3	537.3	Kertindy	449.9	430.3	Martuk	555.8	526.7
Bulaevo	684.6	641.7	Aksu-Ayuly	487	355	Aktobe	509.1	490.4
Vozvyshenka	546.8	610.8	Zharyk	609.7	380.3	Komsomolskoe	358	384.9
Tayinsha	571.4	534.7	Zhana-Arka	356	306.4	Novoalekseevka	437.8	437.7
Chkalovo	538	542.6	Zhetykonyr	134.2	169.1	Novorossiyskoe	586.9	521.5
Kishkenekol	500.1	588.6	Balkash	130.8	164	Karabutak	328.3	348.2
Amangeldi	262.1	287.4	Zhezkazgan	215.3	215.8	Baskuduk	230.4	304.3
Ekiden	260.1	260.6	Bektauata	264.3	229.1	Uil	328.2	302.1
Aralkol	437.2	392	Kyzylzhar	199	258.6	Temir	439.4	406.2
Zheleznodorozhny	427.2	445	Aktogay	226.1	295.8	Nura	191.5	223.4
Karasu	474	475.1	Karaganda	588.1	503.2	Karaulkeldy	314.6	332.6
Presnogorkovka	618.9	621.9	Semiyarka	272.7	291.6	Mugodzharskaya	387.7	368.7
Karabalyk	602.5	661.7	Semipalatinsk	420.8	425.4	Irgiz	185.8	215.8
Michaylovka	535.7	577.7	Shemonaicha	966.2	1066.3	Shalkar	201.4	213.9
Arshalinsky	615	580.4	Leninogorsk	1554.9	880.4	Ayakkum	175.3	174.3
Kostanay	566.8	548	Kainar	318.8	440.1	Taraz	572.6	556
Sarykol	583.3	521.5	Zhalgyz-Tobe	468	591.4	Kulan	557	447.5
Tobol	560.9	529.4	Oskomen	963.7	919.1	Nurlykent	948.7	941.9
Zhetykara	482.8	518.9	Samarka	683.8	731.9	Kordai	961.7	752.9
Kushmurun	431.5	437.7	Ulken-Naryn	678.7	616.3	Ulanbel	166.4	181.3

Stations	Observed	Baseline	Stations	Observed	Baseline	Stations	Observed	Baseline
Dievskaya	418.3	445	Katon-Karagai	841.5	938.7	Anarchai	297.2	602.4
Karamendy	363.9	415.8	Barshatas	258.4	425.4	Otar	544	693.5
Astana	520.3	495.5	Karauyl	327.6	465	Chirik-Rabat	71.7	103.2
Zhaltyr	538.9	462.5	Kokpekty	471.7	553.3	Kyzil-Orda	152.1	152.3
Akkol	675	545.3	Kurshim	364.3	566.8	Shieli	159.5	169.1
Stepnogorsk	515.6	548	Aktogai	229.5	437.7	Kazaly	116.5	116.3
Atbasar	488.6	480.1	Ayagoz	465.3	532	Zhusaly	121.4	128.4
Zhaksy	522.3	477.6	Urzhar	870.6	762	Aral-Tenyzy	132.3	149
Yesil	403	425.4	Aksuat	249.4	583.1	Turkestan	249.6	256.6
Balkashino	708	526.7	Zaisan	524.1	425.4	Tasty	163.4	155.6
Shuchinsk	566	564.1	Backty	443.7	415.8	Shymkent	1380.1	1111.4
Kokshetau	496.6	505.8	Jana Ushtogan	229.3	270.8	Aul T.Ryskulov	1802.3	1236.3
Yereimentau	699.2	537.3	Ganyushkino	160	227.2	Tasaryk	2083.7	1826.2
Korgalzhyn	426.5	411	Karabau	236.3	227.2	Shuyldak	2390.3	1838.9
Golubovka	464.8	513.6	Atyrau	188.6	164	Shardara	307.1	526.7
Zholboldy	425.7	452.5	Kulsary	197.3	154	Kazygurt	977.7	1073.2
Michailovka	490.7	482.7	Akkuduk	128.8	94.8	Ucharal	475.2	435.2
Yertis	446.2	450	Sam	168.9	119.3	Aul #4	124.4	227.2
Fedorovka	427.6	457.5	Beineu	129.9	113.3	Taldy-Korgan	756.5	658.8
Lozovaya	419.4	462.5	Kaztalovka	328.1	413.4	Lepsy	1848.3	542.6
Aktogai	399.6	440.1	Zhalpaktal	377.7	332.6	Kapchagai	427.6	658.8
Uspenka	446	435.2	Zhanybek	469	485.2	Zharkent	242	260.6
Bayanaul	571.6	445	Urda	391.9	380.3	Almaty	1689.2	1336
Ekibastuz	392.7	418.2	Taipak	232.5	260.6	Shelek	375.5	574.9
Evaluation	CC	0.91	**RMSE**	166.6	**NSE**	0.71		

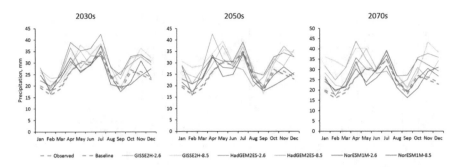

Fig. 5.8 Monthly average precipitation in Kazakhstan in observed, baseline, and different scenarios

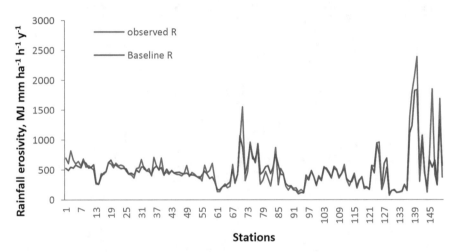

Fig. 5.9 Comparison of the baseline and observed stations for 150 meteorological stations from Kazhydromet

The average annual rainfall erosivity for the observation data ranges from 71.7 to 2390.3 MJ mm ha⁻¹ h⁻¹ year⁻¹ for the Chirik-Rabat and Shuyldak stations, respectively (Table 5.3). The baseline data shows the rainfall erosivity range of 94.8 to 1838.9 MJ mm ha⁻¹ h⁻¹ year⁻¹ in the Akkuduk and Shuyldak stations, respectively. Table 5.3 exemplifies the validation between rainfall erosivity derived from the observation data and that reminded from the base data. The comparison of the observed and baseline R factors yielded 0.81, 0.90, and 156.661 MJ mm ha⁻¹ h⁻¹ year⁻¹ and 0.99 for r², r, and RMSE and NSE, respectively, indicating an excellent model performance for rainfall erosivity prediction for the current and future climates.

The predicted annual rainfall erosivity values from the three GCM projections were compared with those calculated using the gridded monthly rainfall data obtained from WorldClim for the baseline period (1960–1990). The rainfall erosivity values derived from the three GCM ensembles vary across the RCP 2.6 and RCP 8.5 compared with the results from the gridded baseline data. The results did not exhibit a particular trend with regard to increase or decrease as the period increases;

instead, the trend is synonymous with that in the average annual precipitation. Spatially, for all GCMs and baseline climate, rainfall erosivity is higher to the east of the country but lower to the west.

Table 5.4 presents the impact of precipitation on the current and future rainfall erosivities in Kazakhstan. The GSM ensembles show that precipitation and rainfall erosivity significantly increase from the baseline across all the ensembles. The average of all the ensembles indicates that precipitation will increase during the 2030s, 2050s and 2070s to 285.5 mm, 290.2 mm, and 302 mm, respectively, from baseline (250.6 mm). However, the GISSE2H has projected a higher increase in precipitation than the rest of the ensembles with the same scenario and period. There will also be an average increase in rainfall erosivity for the 2030s, 2050s, and 2070s of about 140.8, 187, and 278.5 MJ mm ha^{-1} h^{-1} year^{-1} respectively.

Precipitation and erosivity vary across individual GCM without exhibiting a steady increase or decrease, whereas the average annual precipitation and erosivity exhibit a steady rise across all the ensembles put together (Table 5.5). Even though the average annual rainfall and erosivity exhibit steady increases in all the climate models combined from the baseline climate, only NorESM1M-8.5 reveals a likely decrease in the annual rainfall and erosivity in the 2050s.

Table 5.4 Changes in the average annual rainfall erosivity (MJ mm ha^{-1} h^{-1} year^{-1}) under past and future climate changes across Kazakhstan

Periods	Projections	Precipitation	Rainfall erosivity	Change, %	Erosivity density	Change, %
	Baseline	250.57	374.02	0.00	1.38	0.00
2030s	GISSE2H-2.6	307.00	514.79	37.64	1.57	13.43
	GISSE2H-8.5	305.38	509.38	36.19	1.48	7.15
	HadGEM2ES-2.6	320.12	550.52	47.19	1.61	16.30
	HadGEM2ES-8.5	287.71	468.32	25.21	1.50	8.67
	NorESM1M-2.6	266.51	407.90	9.06	1.44	4.18
	NorESM1M-8.5	261.24	397.06	6.16	1.42	2.77
	Average	291.33	474.66	26.91	1.50	8.75
2050s	GISSE2H-2.6	303.43	503.68	34.67	1.56	12.73
	GISSE2H-8.5	323.34	560.97	49.98	1.62	17.00
	HadGEM2ES-2.6	302.30	503.48	34.61	1.55	12.22
	HadGEM2ES-8.5	300.76	498.75	33.35	1.38	-0.25
	NorESM1M-2.6	262.05	398.92	6.66	1.42	2.98
	NorESM1M-8.5	249.55	371.88	-0.57	1.38	-0.25
	Average	290.24	472.95	26.45	1.49	7.40
2070s	GISSE2H-2.6	305.16	508.81	36.04	1.56	13.06
	GISSE2H-8.5	354.60	652.49	74.45	1.56	13.06
	HadGEM2ES-2.6	302.46	504.01	34.75	1.55	12.25
	HadGEM2ES-8.5	326.93	570.52	52.54	1.63	17.76
	NorESM1M-2.6	267.94	414.22	10.75	1.44	4.31
	NorESM1M-8.5	254.71	384.22	2.73	1.40	0.99
	Average	301.97	505.71	35.21	1.52	10.24

Table 5.5 Rainfall erosivity in Kazakhstan by regions

Periods	Projections	Almaty	Akmola	Aktobe	Atyrau	East Kazakhstan	Mangystau	North Kazakhstan	Pavlodar	Karagandy	Kostanay	Kyzylorda	South Kazakhstan	West Kazakhstan	Zhambyl
	Baseline	535	484	299	205	635	118	562	424	283	425	142	461	387	346
2030s	GISSE2H-2.6	716	638	414	295	889	171	690	551	441	552	218	609	477	518
	Change, %	34	32	39	44	40	46	23	30	56	30	53	32	23	50
	GISSE2H-8.5	691	650	402	267	830	169	764	581	446	580	230	576	471	472
	Change, %	29	34	35	30	31	43	36	37	58	37	62	25	22	36
	HadGEM2ES-2.6	767	766	450	300	820	164	940	634	434	678	218	686	490	539
	Change, %	43	58	51	46	29	39	67	50	54	60	53	49	26	56
	HadGEM2ES-8.5	637	630	355	268	797	118	784	553	364	566	161	537	473	416
	Change, %	19	30	19	31	25	0	40	31	29	33	13	17	22	20
	NorESM1M-2.6	567	510	369	244	616	161	609	407	309	507	183	505	416	380
	Change, %	6	5	23	19	-3	37	8	-4	9	19	29	10	7	10
	NorESM1M-8.5	588	504	335	204	646	127	549	410	327	464	164	486	365	384
	Change, %	10	4	12	0	2	8	-2	-3	16	9	15	6	-6	11
2050s	GISSE2H-2.6	721	599	394	308	857	180	666	514	422	539	228	636	463	544
	Change, %	35	24	32	50	35	53	18	21	49	27	61	38	19	57
	GISSE2H-8.5	817	686	434	300	966	180	789	602	476	619	235	644	525	559
	Change, %	53	42	45	46	52	53	40	42	68	46	65	40	36	62
	HadGEM2ES-2.6	707	671	394	303	781	151	845	580	393	618	185	618	482	485
	Change, %	32	38	32	47	23	28	50	37	39	45	30	34	24	40
	HadGEM2ES-8.5	693	652	378	296	793	140	807	541	405	597	192	643	485	493
	Change, %	30	35	26	44	25	19	44	28	43	40	35	40	25	43
	NorESM1M-2.6	582	502	349	228	627	149	593	399	296	483	167	491	409	376
	Change, %	9	4	17	11	-1	27	6	-6	5	14	17	7	6	9
	NorESM1M-8.5	571	451	285	201	651	119	525	397	286	424	143	468	367	353
	Change, %	7	-7	-5	-2	2	1	-7	-6	1	0	0	2	-5	2

(continued)

Table 5.5 (continued)

Periods	Projections	Almaty	Akmola	Aktobe	Atyrau	East Kazakhstan	Mangystau	North Kazakhstan	Pavlodar	Karagandy	Kostanay	Kyzylorda	South Kazakhstan	West Kazakhstan	Zhambyl
2070s	GISSE2H-2.6	690	634	433	302	830	178	741	557	411	582	217	609	481	507
	Change, %	29	31	45	47	31	51	32	31	46	37	53	32	24	46
	GISSE2H-8.5	904	836	524	341	1127	185	930	731	576	747	274	705	530	637
	Change, %	69	73	75	66	78	57	65	73	104	76	92	53	37	84
	HadGEM2ES-2.6	724	626	396	314	809	141	775	550	409	596	187	620	505	498
	Change, %	36	29	33	53	27	19	38	30	45	40	32	35	30	44
	HadGEM2ES-8.5	794	719	428	333	943	167	896	616	457	664	219	773	542	590
	Change, %	48	48	43	62	49	42	60	45	62	56	54	68	40	70
	NorESM1M-2.6	558	537	372	257	657	147	645	440	306	520	155	463	452	363
	Change, %	4	11	25	25	3	25	15	4	8	22	9	1	17	5
	NorESM1M-8.5	565	500	312	209	640	133	593	415	279	469	145	460	380	355
	Change, %	6	3	4	2	1	13	5	−2	−1	10	2	0	−2	3

Mean baseline and estimated rainfall erosivity (MJ mm ha^{-1} h^{-1} year^{-1}) by GISSE2H, HadGEM2ES, and NorESM1M with RCP 2.6 and RCP 8.5 emission scenario models. Projected change concerning baseline (%)

The implication of change in erosivity due to the current and future changes in climate is presented in Tables 5.4 and 5.5, indicating variations in the potential annual erosivity for the current climate and future scenarios. Figure 5.11 presents the spatial variability of the projected rainfall erosivity and erosivity density for Kazakhstan. In addition, Fig. 5.12 presents the relative difference between each projected scenario and the models' baseline. The highest values occur in the East Kazakhstan region, and the lowest values are in the Mangystau region for the observed data and all models (Table 5.5).

5.4   Erosivity Density in Kazakhstan

The separately projected annual erosivity results (Fig. 5.10) are divided by a corresponding average rainfall data subsequent in the average erosivity density ratio. The annual erosivity density for the baseline period has a mean value of 1.38 MJ ha^{-1}

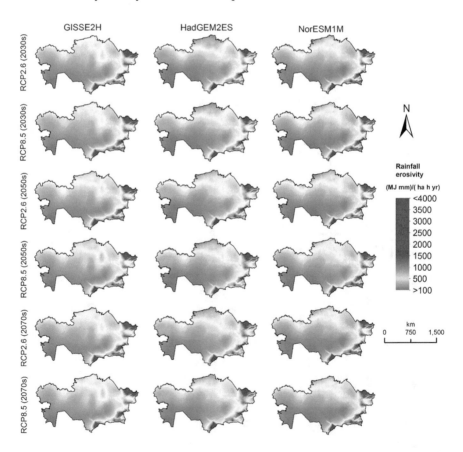

Fig. 5.10 Rainfall erosivity projections for the years 2030s, 2050s, and 2070s according to RCP 2.6 and RCP 8.5 scenarios driven by the GISSE2H, HadGEM2ES, and NorESM1M GCM models in Kazakhstan

h⁻¹, with variability ranging from 0.62 to 3.69 MJ ha⁻¹ h⁻¹. The projected variability of erosivity density is also very high, as the HadGEM2ES-2.6 has the highest mean erosivity densities of 1.61, 1.55, and 1.63 MJ ha⁻¹ h⁻¹ in the 2030s, 2050s, and 2070s, respectively, followed by GISSE2H-2.6 with erosivity densities of 1.57, 1.56, and 1.56 MJ ha⁻¹ h⁻¹ in the 2030s, 2050s, and 2070s, respectively. The erosivity densities of BCCCSM1.1-2.6 and BCCCSM1.1-8.5 are 1.45 and 1.43 in the 2070s (Table 5.4, Figs. 5.11 and 5.12).

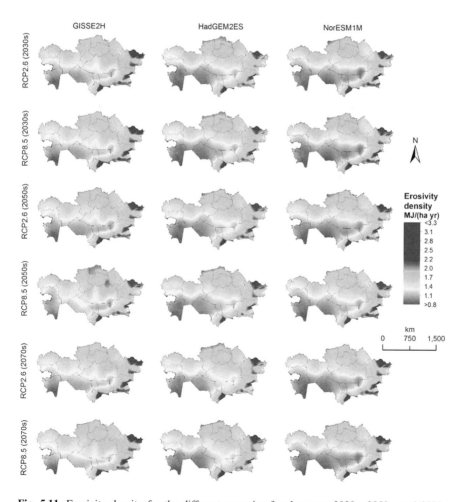

Fig. 5.11 Erosivity density for the different scenarios for the years 2030s, 2050s, and 2070s according to RCP 2.6 and RCP 8.5 scenarios driven by the GISSE2H, HadGEM2ES, and NorESM1M GCM models in Kazakhstan

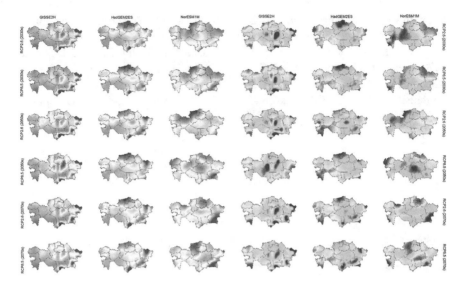

Fig. 5.12 Absolute differences of rainfall erosivity and erosivity density in the 2030, 2050, and 2070 projections in Kazakhstan

Chapter 6
Conclusions and Recommendations

Abstract This chapter concludes the preceding chapters and offers recommendations for carrying out future studies. Higher amounts of rainfall were the main factor for the spatiotemporal variability in rainfall erosivity. Public policies aimed at preserving soil and water resources should be encouraged and applied at the national land survey level. Further study is required to consider other essential influences that intensify the erosivity and erosion, particularly the future land cover changes.

Keywords Central Asia · Rainfall erosivity · Soil erosion · Erosivity density · Climate change · RUSLE · Recommendations · Spatiotemporal variability

It has been shown that Central Asia is a zone sensitive to climate change. A quantitative assessment of the impact of climate change and its impact on soil erosion over Central Asia is important to assist policymakers and land managers in adopting the strategies for their protection and conservation. However, the reported quantitative observations of soil erosion in Central Asia are limited. The goal of the study was to use globally available data to estimate the spatial and temporal variations and prediction of climate-induced rainfall erosivity and soil erosion in regions with limited data, such as Central Asia. The study was conducted in Central Asia. The sections below discuss the obtained results.

In this research, we estimated the potential influence of climate change on rainfall erosivity and erosivity density in Central Asia using baseline data (1960–1990) and projected precipitation data (2020–2049 and 2060–2089). The projected precipitation was obtained from the downscaled data of four GCMs, BCCCSM1.1, IPSLCM5ALR, MIROC5, and MPIESMLR, based on two scenarios, RCP2.6 and RCP8.5. The mean erosion of rainfall in Central Asia was calculated and compared with the climate scenario predictions. The mean rainfall erosivity within the baseline period was 497.8 MJ mm ha^{-1} h^{-1} year^{-1}.

E. Duulatov et al., *Current and Future Trends of Rainfall Erosivity and Soil Erosion in Central Asia*, SpringerBriefs in Environmental Science,
https://doi.org/10.1007/978-3-030-63509-1_6

Tajikistan and Kyrgyzstan are predicted to be the countries most affected by rainfall erosivity. The increasing trends in annual rainfall erosivity from baseline climate up to the GCMs and the climate scenarios experienced variations in rainfall erosivity. There is a positive change in the average annual rainfall erosivity of 5.6% and 9.6% in the 2030s and 2070s, respectively, compared with the baseline (1960–1990).

The BCCCSM1.1 scenarios projected both an increase and a decrease in the mean rainfall erosivity in Kyrgyzstan (−7.8% to 6.3%), Tajikistan (−26.3% to 3.8%), Turkmenistan (−19% to 10.5%), and Uzbekistan (−17.7% to 9%) and an increase in Kazakhstan (11.6% to 23.2%). The IPSLCM5ALR scenarios projected a decrease in the mean rainfall erosivity in Kyrgyzstan (−14.5% to −35.5%), Tajikistan (−8.2% to −26.2%), Turkmenistan (−7.1% to −16.7%), and Uzbekistan (−7.7% to −28.2%) and both an increase and a decrease in Kazakhstan (−1.6% to 1.5%). The MIROC5 scenarios projected an increase in Kyrgyzstan (27.9% to 50.1%), Kazakhstan (5.3% to 27.3%), Tajikistan (34.3% to 56.3%), and Uzbekistan (2.1% to 24.1%) and both an increase and a decrease in Turkmenistan (−4.7% to 19.6%). The MPIESMLR scenarios projected a mean rainfall erosivity increase in Kyrgyzstan (0.9% to 27%), Kazakhstan (0.6% to 15.7%), and Turkmenistan (2.3% to 14.3%) and both an increase and decrease in Tajikistan (−2.2% to 24.4%) and Uzbekistan (−2.7% to 15.3%). The average values of erosion variations presented in this study are average changes in countries, whereas within these countries, we found both an increase and a decrease, which emphasize some spatial variabilities of rainfall erosivity and soil erosion in Central Asia.

The aggregate average annual precipitation and erosion activity for all climate models for all scenarios exhibit a steady growth compared with the baseline climate. Only IPSLCM5ALR (RCP 2.6 and RCP 8.5) shows a decrease in the average erosivity for the 2030 and 2070 scenarios. Higher amounts of rainfall were the main factor for the spatiotemporal variability in rainfall erosivity.

The spatial–temporal variation of rainfall erosivity in Kazakhstan in 1970–2017 was investigated by the model of Renard and Freimund (1994). The results revealed that the average annual rainfall erosivity in Kazakhstan over the past 48 years was 464 MJ mm ha^{-1} h^{-1} year^{-1}. No significant time trend was observed in the annual rainfall erosivity. Some of the results presented here are relevant to the further study of potential soil erosion in Kazakhstan.

The East Kazakhstan, North Kazakhstan, and Almaty regions were under a more significant threat of rainfall erosivity than other regions. Most of the meteorological stations observed exhibited non-significant upward trends and few downward trends. The rainfall erosivity in Kazakhstan exhibits a tendency to increase, and the probability of water and soil erosion caused by precipitation is continuously increasing.

This study estimated the long-term variations in the annual rainfall erosivity in Kazakhstan using past and future climatic data. The increasing trends in the annual rainfall erosivity were obtained from the baseline climate up to the GCMs (GISSE2H, HadGEM2-ES, and NorESM1M), and the climate scenarios experienced variations in rainfall erosivity. Based on the baseline climate, the average percentages of change in rainfall erosivity are 26.9%, 26.4%, and 35.2% in the 2030s, 2050s, and 2070s, respectively. The aggregate average annual precipitation and erosion activity for all climate models for all scenarios exhibit a steady growth compared with the baseline climate.

Based on the obtained results, the following achievements are presented:

- The increasing and decreasing rates of rainfall erosivity and erosivity density were analyzed and validated using four GCMs. Out of the four GCMs, only IPSLCM5ALR-2.6 and RCP 8.5 exhibit a decrease in the annual rainfall and erosivity in the 2030s and 2070s, respectively, in Central Asia.
- The spatial–temporal variation of rainfall erosivity in Kazakhstan in 1970–2017 was investigated. No significant time trend was observed in the annual rainfall erosivity. The long-term variations in the annual rainfall erosivity in Kazakhstan was estimated using past and future climatic data.

The result will facilitate in appropriately preventing erosion in areas with severe consequences. In addition, the results of this study can help in the development of management scenarios and enable policymakers to effectively manage the risks of soil erosion to prioritize different regions for an adequate policy implementation. Moreover, the results of this study may be general and reflect the type of changes that have occurred or are expected in other arid and semi-arid areas. This study examined erosion caused by rainfall and water. Other types of erosion, such as wind erosion (typical in many regions in Central Asia) and erosion from freeze–thaw, are not considered due to the lack of relevant data. More research is required to include other important factors contributing to the future increase in soil erosion, especially future changes in the land cover.

Recommendations

- The potential soil erosion map will help define the plots that are most vulnerable to soil erosion.
- The scaled vision of soil erosion will allow the formation of science-based recommendations for soil conservation and help establish measures based on advanced international experience (multiple cropping, strip cropping, mulching, terracing, contour plowing, and other ways of soil conservation).
- Local executive bodies and authorities should implement soil conservation recommendations in situ and according to legislation.
- To prevent soil erosion on slopes, conservation measures, such as increased forestation and shrub planting, should be considered.
- The application of GIS and RS technologies allows the assessment of soil erosion. In future research, it would be helpful to focus on particular plots, which are more prone to soil erosion, by applying data in higher resolution.

Appendix

Average annual rainfall, rainfall erosivity, and erosivity density in Kazakhstan (1970–2017)

No	Stations	Lat (∘)	Long (∘)	Altitude (m)	Annual rainfall (mm)	R factor (MJ mm ha^{-1} h^{-1} year^{-1})	Erosivity density (MJ ha^{-1} h^{-1})
1	Petropavl	53.49	69.08	262	384.6	701.3	1.82
2	Ruzaevka	52.8	67	233	352.4	609.2	1.73
3	Saumalkol	53.3	68.1	310	423.6	819.2	1.93
4	Sergeevka	53.9	67.4	143	371.6	663.5	1.79
5	Blagoveshenka	54.4	67	138	349.5	600.9	1.72
6	Yavlenka	54.4	68.4	116	363.9	641.4	1.76
7	Timiryazevo	53.7	66.4	165	336.5	565.3	1.68
8	Bulaevo	54.9	70.5	128	378.9	684.6	1.81
9	Vozvyshenka	54.4	70.9	123	329.6	546.8	1.66
10	Tayinsha	53.8	69.8	164	338.7	571.4	1.69
11	Chkalovo	53.6	70.5	159	326.2	538.0	1.65
12	Kishkenekol	53.6	72.4	124	311.8	500.1	1.60
13	Amangeldy	50.2	65.18	137	208.7	262.1	1.26
14	Ekidyn	49.53	66.13	202	207.7	260.1	1.25
15	Aralkol	51.1	62.9	254	286.8	437.2	1.52
16	Zheleznodorozhny	52.1	65.6	254	282.7	427.2	1.51
17	Karasu	52.7	65.5	202	301.6	474.0	1.57
18	Presnogorkovka	54.5	65.8	159	355.9	618.9	1.74
19	Karabalyk	53.8	62.1	182	350.0	602.5	1.72
20	Michailovka	53.6	64.3	185	325.4	535.7	1.65
21	Arshalinsky	52.7	61.2	280	354.5	615.0	1.73
22	Kostanay	53.2	63.6	167	337.0	566.8	1.68

E. Duulatov et al., *Current and Future Trends of Rainfall Erosivity and Soil Erosion in Central Asia*, SpringerBriefs in Environmental Science, https://doi.org/10.1007/978-3-030-63509-1

No	Stations	Lat (°)	Long (°)	Altitude (m)	Annual rainfall (mm)	R factor (MJ mm ha⁻¹ h⁻¹ year⁻¹)	Erosivity density (MJ ha⁻¹ h⁻¹)
23	Sarykol	53.3	65.6	197	343.0	583.3	1.70
24	Tobol	52.7	62.6	203	334.8	560.9	1.68
25	Zhetykara	52.2	61.2	267	305.0	482.8	1.58
26	Kushmurun	52.4	64.6	109	284.5	431.5	1.52
27	Dievskaya	52	63.7	213	279.1	418.3	1.50
28	Karamendy	51.7	64.3	217	255.9	363.9	1.42
29	Astana	51.2	71.4	352	319.6	520.3	1.63
30	Zhaltyr	51.6	69.8	304	326.6	538.9	1.65
31	Akkol	52	70.9	400	375.6	675.0	1.80
32	Stepnogorsk	52.4	71.8	283	317.8	515.6	1.62
33	Atbasar	51.8	68.4	276	307.3	488.6	1.59
34	Zhaksy	51.9	67.3	380	320.3	522.3	1.63
35	Yesil	51.9	66.3	217	272.7	403.0	1.48
36	Balkashino	52.5	68.8	396	386.9	708.0	1.83
37	Shuchinsk	52.9	70.2	394	336.7	566.0	1.68
38	Kokshetau	53.3	69.4	228	310.4	496.6	1.60
39	Yereimentau	51.6	73.1	405	383.9	699.2	1.82
40	Korgalzhyn	50.6	70	326	282.4	426.5	1.51
41	Golubovka	53.1	74.2	116	297.9	464.8	1.56
42	Zholboldy	52.7	74.9	120	282.1	425.7	1.51
43	Michailovka	53.8	76.5	103	308.1	490.7	1.59
44	Yertis	53.4	75.5	85	290.5	446.2	1.54
45	Fedorovka	53.4	76.3	106	282.9	427.6	1.51
46	Lozovaya	53.3	77.9		279.5	419.4	1.50
47	Aktogai	53	76	102	271.2	399.6	1.47
48	Uspenka	52.9	77.4	113	290.4	446.0	1.54
49	Bayanaul	50.8	75.7	531	338.8	571.6	1.69
50	Ekibastuz	51.8	75.4	181	268.3	392.7	1.46
51	Pavlodar	52.3	76.9	106	293.9	454.7	1.55
52	Sharbakty	52.4	78.2	147	285.0	432.8	1.52
53	Shalday	51.9	78.8	156	265.1	385.1	1.45
54	Koktobe	51.5	77.3	134	266.3	387.9	1.46
55	Besoba	49.4	74.5	710	233.6	314.2	1.34
56	Korneevka	50.2	74.2	595	340.7	576.8	1.69
57	Kertindy	49.9	71.8	493	292.0	449.9	1.54
58	Aksu-Ayuly	48.8	73.7	740	306.7	487.0	1.59
59	Zharyk	48.9	72.9	700	352.6	609.7	1.73
60	Zhana-Arka	48.7	71.7	488	252.4	356.0	1.41
61	Zhetykonyr	46.7	68.3	272	137.8	134.2	0.97
62	Balkash	46.8	75.1	341	135.6	130.8	0.97
63	Zhezkazgan	47.8	67.7	371	184.7	215.3	1.17
64	Bektauata	47.5	74.8	740	209.8	264.3	1.26

No	Stations	Lat (°)	Long (°)	Altitude (m)	Annual rainfall (mm)	R factor (MJ mm ha^{-1} h^{-1} year^{-1})	Erosivity density (MJ ha^{-1} h^{-1})
65	Kzylzhar	48.3	69.7	361	175.9	199.0	1.13
66	Aktogay	48.2	75.1	745	190.4	226.1	1.19
67	Karaganda	50.2	72.8	519	344.8	588.1	1.71
68	Semiyarka	50.86	78.35	144	213.9	272.7	1.27
69	Semipalatinsk	50.35	80.25	205	280.1	420.8	1.50
70	Shemonaicha	50.63	81.91	314	469.4	966.2	2.06
71	Leninogorsk	49.7	82.66	466	630.7	1554.9	2.47
72	Kainar	49.2	77.4	826	235.7	318.8	1.35
73	Zhalgyz-Tobe	49.21	81.21	455	299.2	468.0	1.56
74	Ust-Kamenogorsk	50.03	82.5	285	468.6	963.7	2.06
75	Samarka	49.03	83.4	612	378.7	683.8	1.81
76	Ulken-Naryn	49.2	84.51	393	376.9	678.7	1.80
77	Katon-Karagai	49.13	85.65	1751	430.8	841.5	1.95
78	Barshatas	48.16	78.41	679	206.9	258.4	1.25
79	Karauyl	48.91	79.2	678	239.7	327.6	1.37
80	Kokpekty	48.75	82.36	539	300.7	471.7	1.57
81	Kurshim	48.58	83.65	421	256.1	364.3	1.42
82	Aktogai	46.93	79.65	361	192.2	229.5	1.19
83	Ayagoz	47.93	80.45	653	298.1	465.3	1.56
84	Urzhar	47.11	81.61	486	439.9	870.6	1.98
85	Aksuat	47.78	82.66	580	202.4	249.4	1.23
86	Zaisan	47.46	84.91	582	321.0	524.1	1.63
87	Backty	46.65	82.75	443	289.5	443.7	1.53
88	Novyi Ushtogan	47.9	48.8	−13	192.1	229.3	1.19
89	Ganyushkino	46.63	49.13	−27	153.6	160.0	1.04
90	Karabau	48.45	52.91	−4	195.7	236.3	1.21
91	Atyrau	47.01	51.85	−24	170.2	188.6	1.11
92	Kulsary	46.8	53.91	−17	175.0	197.3	1.13
93	Akkuduk	42.96	54.11	74	134.3	128.8	0.96
94	Sam	45.4	56.11	82	158.9	168.9	1.06
95	Beineu	45.33	55.2	4	135.0	129.9	0.96
96	Kaztalovka	49.77	48.7	14	240.0	328.1	1.37
97	Zhalpaktal	49.66	49.48	7	261.9	377.7	1.44
98	Zhanybek	49.4	46.8		299.6	469.0	1.57
99	Urda	48.75	47.43	2	268.0	391.9	1.46
100	Taipak	49.05	51.86	−11	193.7	232.5	1.20
101	Chapaevo	50.2	51.16	7	271.0	399.0	1.47
102	Karatyuba	49.68	53.51	40	243.8	336.6	1.38
103	Uralsk	51.25	51.4	28	332.3	554.0	1.67
104	Aksai	51.18	53	59	315.3	509.1	1.61
105	Chingirlau	51.03	54.1	74	286.3	435.9	1.52
106	Zhambeity	50.25	52.56	28	253.5	358.3	1.41

No	Stations	Lat (°)	Long (°)	Altitude (m)	Annual rainfall (mm)	R factor (MJ mm ha⁻¹ h⁻¹ year⁻¹)	Erosivity density (MJ ha⁻¹ h⁻¹)
107	Martuk	50.75	56.53	177	332.9	555.8	1.67
108	Aktobe	50.28	57.15	220	315.3	509.1	1.61
109	Komsomolskoe	50.41	60.46	278	253.3	358.0	1.41
110	Novoalekseevka	50.2	55.6	134	287.0	437.8	1.53
111	Novorossiyskoe	50.23	58	418	344.4	586.9	1.70
112	Karabutak	49.95	60.13	225	240.1	328.3	1.37
113	Baskuduk	49.75	61.43	217	192.7	230.4	1.20
114	Uil	49.06	54.68	72	240.0	328.2	1.37
115	Temir	49.15	57.11	238	287.7	439.4	1.53
116	Nura	48.83	62.1	82	171.7	191.5	1.11
117	Karaulkeldy	48.73	56.03	202	233.8	314.6	1.35
118	Mugodzharskaya	48.63	58.5	426	266.2	387.7	1.46
119	Irgiz	48.61	61.26	100	168.6	185.8	1.10
120	Shalkar	47.85	59.61	169	177.2	201.4	1.14
121	Ayakkum	46.66	59	135	162.6	175.3	1.08
122	Taraz	42.85	71.38	649	339.1	572.6	1.69
123	Kulan	42.95	72.75	682	333.4	557.0	1.67
124	Nurlykent	42.7	70.81	967	464.1	948.7	2.04
125	Kordai	43.31	74.95	1140	468.0	961.7	2.05
126	Ulanbel	44.8	71.06	270	157.4	166.4	1.06
127	Anarchai	44.01	75.25	960	225.7	297.2	1.32
128	Otar	43.53	75.25	735	328.5	544.0	1.66
129	Chirik-Rabat	44.06	62.9	84	93.3	71.7	0.77
130	Kyzil-Orda	44.76	65.53	124	148.9	152.1	1.02
131	Shieli	44.16	66.75	154	153.3	159.5	1.04
132	Kazaly	45.76	62.11	67	126.1	116.5	0.92
133	Zhusaly	45.51	64.08	99	129.4	121.4	0.94
134	Aral-Tenyzy	46.78	61.65	53	136.5	132.3	0.97
135	Turkestan	43.26	68.21	203	202.5	249.6	1.23
136	Tasty	44.49	69.14	244	155.7	163.4	1.05
137	Shymkent	42.3	69.6	516	585.7	1380.1	2.36
138	Aul Turar Ryskulov	42.48	70.3	864	691.3	1802.3	2.61
139	Tasaryk	42.23	70.15	1134	756.5	2083.7	2.75
140	Shuyldak	42.3	70.43	1994	823.8	2390.3	2.90
141	Shardara	41.33	67.91	244	230.3	307.1	1.33
142	Kazygurt	41.75	69.36	633	472.8	977.7	2.07
143	Ucharal	46.16	80.93	392	302.0	475.2	1.57
144	Aul N 4	45.46	75.21	360	131.4	124.4	0.95
145	Taldy-Korgan	45.01	78.38	603	403.2	756.5	1.88
146	Lepsy	45.53	80.61	994	702.2	1848.3	2.63
147	Kapchagai	44.03	77.08	695	282.9	427.6	1.51

No	Stations	Lat (°)	Long (°)	Altitude (m)	Annual rainfall (mm)	R factor (MJ mm ha^{-1} h^{-1} year^{-1})	Erosivity density (MJ ha^{-1} h^{-1})
148	Zharkent	44.16	80.06	637	198.6	242.0	1.22
149	Almaty	43.23	76.93	871	664.0	1689.2	2.54
150	Shelek	43.6	78.25	598	261.0	375.5	1.44

References

Aizen VB, Kuzmichenok VA, Surazakov AB, Aizen EM (2007) Glacier changes in the Tien Shan as determined from topographic and remotely sensed data. Glob Planet Chang 56:328–340

Alamanov S, Lelevkin V, Podrezov O, Podrezov A (2006) Climate changes and water problems in Central Asia. United Nations Environment Program (UNEP) and World Wildlife Fund (WWF), Moscow-Bishkek (in Russian)

Alewell C, Borrelli P, Meusburger K, Panagos P (2019) Using the USLE: Chances, challenges and limitations of soil erosion modelling. Int Soil Water Conser Res 7:203–225. https://doi.org/10.1016/j.iswcr.2019.05.004

Almagro A, Oliveira PTS, Nearing MA, Hagemann S (2017) Projected climate change impacts in rainfall erosivity over Brazil. Sci Rep 7:8130. https://doi.org/10.1038/s41598-017-08298-y

Almagro A, Thomé TC, Colman CB, Pereira RB, Marcato Junior J, Rodrigues DBB, Oliveira PTS (2019) Improving cover and management factor (C-factor) estimation using remote sensing approaches for tropical regions. Int Soil Water Conser Res 7:325–334. https://doi.org/10.1016/j.iswcr.2019.08.005

Amanambu AC, Li L, Egbinola CN, Obarein OA, Mupenzi C, Chen D (2019) Spatio-temporal variation in rainfall-runoff erosivity due to climate change in the Lower Niger Basin, West Africa. CATENA 172:324–334

Angulo-Martínez M, Beguería S (2009) Estimating rainfall erosivity from daily precipitation records: A comparison among methods using data from the Ebro Basin (NE Spain). J Hydrol 379:111–121

Arnold JG, Srinivasan R, Muttiah RS, Williams JR (1998) Large area hydrologic modeling and assessment part I: model development. JAWRA J Am Water Res Assoc 34:73–89

Arnoldus H (1977) Methodology used to determine the maximum potential average annual soil loss due to sheet and rill erosion in Morocco. FAO Soil Bull

Arnoldus H (1980) An approximation of the rainfall factor in the Universal Soil Loss Equation. In: De Boodt M, Gabriels D (eds) Assessment of Erosion. Wiley, New York, pp 127–132

Batjes N (1997) Methodological framework for assessment and mapping of the vulnerability of soils to diffuse pollution at a continental level (SOVEUR project). International Soil Reference and Information Centre, Wageningen

Beasley D, Huggins L, Monke A (1980) ANSWERS: A model for watershed planning. Trans ASAE 23:938–0944

Benavidez R, Jackson B, Maxwell D, Norton K (2018) A review of the (Revised) Universal Soil Loss Equation ((R) USLE): with a view to increasing its global applicability and improving soil loss estimates. Hydrol Earth Syst Sci 22:6059–6086

The Author(s), under exclusive license to Springer Nature Switzerland AG 2021

E. Duulatov et al., *Current and Future Trends of Rainfall Erosivity and Soil Erosion in Central Asia*, SpringerBriefs in Environmental Science, https://doi.org/10.1007/978-3-030-63509-1

Borrelli P et al (2017) An assessment of the global impact of 21st century land use change on soil erosion. Nat Commun 8:2013. https://doi.org/10.1038/s41467-017-02142-7

Borrelli P, Meusburger K, Ballabio C, Panagos P, Alewell C (2018) Object-oriented soil erosion modelling: A possible paradigm shift from potential to actual risk assessments in agricultural environments. Land Degrad Dev 29:1270–1281

Borrelli P et al (2020) Land use and climate change impacts on global soil erosion by water (2015-2070). Proc Natl Acad Sci 202001403. https://doi.org/10.1073/pnas.2001403117

Briggs D, Giordano A, Cornaert M, Peter D, Maef J (1992) CORINE soil erosion risk and important land resources in the southern regions of the European community. EUR

Campbell JL, Driscoll CT, Pourmokhtarian A, Hayhoe K (2011) Streamflow responses to past and projected future changes in climate at the Hubbard Brook Experimental Forest, New Hampshire, United States. Water Resour Res 47

Carter JG, Cavan G, Connelly A, Guy S, Handley J, Kazmierczak A (2015) Climate change and the city: Building capacity for urban adaptation. Prog Plan 95:1–66

Change IC (2014) Impacts, adaptation, and vulnerability. part B: regional aspects. Contribution of Working Group II to the Fifth Assessment Report of the Intergovernmental Panel on Climate Change. Cambridge Univesity Press, Cambridge, UK/New York

Chen X, Zhou Q (2015) Ecological and Environmental Remote Sensing in Arid Zone–A Case Study of Central Asia. Science Press, Beijing

Chen X, Wang S, Hu Z, Zhou Q, Hu Q (2018) Spatiotemporal characteristics of seasonal precipitation and their relationships with ENSO in Central Asia during 1901–2013. J Geogr Sci 28:1341–1368

Chevallier P, Pouyaud B, Mojaïsky M, Bolgov M, Olsson O, Bauer M, Froebrich J (2014) River flow regime and snow cover of the Pamir Alay (Central Asia) in a changing climate. Hydrol Sci J 59:1491–1506. https://doi.org/10.1080/02626667.2013.838004

Colman C (2018) Impacts of climate and land use changes on soil erosion in the Upper Paraguay Basin MSc Dissertação. Federal University of Mato Grosso do Sul, Campo Grande, MS, Brazil

Dai A, Wigley T (2000) Global patterns of ENSO-induced precipitation. Geophys Res Lett 27:1283–1286

Denisov VV (2006) Ecology. Moscow-Rostov na Donu

Desmet P, Govers G (1996) A GIS procedure for automatically calculating the USLE LS factor on topographically complex landscape units. J Soil Water Conserv 51:427–433

Dobrovolsky GV (2002) Degradation and protection of soils. Moscow State University Publishing House, Moscow

Duishonakunov M, Imbery S, Narama C, Mohanty A, King L (2014) Recent glacier changes and their impact on water resources in Chon and Kichi Naryn Catchments. Kyrgyz Rep Water Sci Technol Water Supp 14:444–452

Durigon V, Carvalho D, Antunes M, Oliveira P, Fernandes M (2014) NDVI time series for monitoring RUSLE cover management factor in a tropical watershed. Int J Remote Sens 35:441–453

Duulatov E, Chen X, Amanambu AC, Ochege FU, Orozbaev R, Issanova G, Omurakunova G (2019) Projected Rainfall Erosivity Over Central Asia Based on CMIP5 Climate Models. Water 11:897

Flanagan D, Nearing M (1995) USDA-Water Erosion Prediction Project: Hillslope profile and watershed model documentation. NSERL report

Foster G, Meyer L (1977) Soil erosion and sedimentation by water – an overview. In: ASAE Publication No. 4-77. Proceedings of the national symposium on soil erosion and sediment by water, Chicago, IL, December 12–13, 1977

Gafforov KS et al (2020) The Assessment of Climate Change on Rainfall-Runoff Erosivity in the Chirchik–Akhangaran Basin. Uzbekistan Sustain 12:3369

Gafurov A (2010) Water balance modeling using remote sensing information: focus on Central Asia

Ganasri B, Ramesh H (2016) Assessment of soil erosion by RUSLE model using remote sensing and GIS-A case study of Nethravathi Basin. Geosci Front 7:953–961

Gaubi I, Chaabani A, Mammou AB, Hamza M (2017) A GIS-based soil erosion prediction using the Revised Universal Soil Loss Equation (RUSLE)(Lebna watershed, Cap Bon, Tunisia). Nat Hazards 86:219–239

Gedefaw M, Yan D, Wang H, Qin T, Girma A, Abiyu A, Batsuren D (2018) Innovative Trend Analysis of Annual and Seasonal Rainfall Variability in Amhara Regional State. Ethiopia Atmos 9:326

Gelagay H (2016) RUSLE and SDR Model Based Sediment Yield Assessment in a GIS and Remote Sensing Environment; A Case Study of Koga Watershed, Upper Blue Nile Basin. Ethiopia Hydrol Curr Res 7:2

Gu C, Mu X, Gao P, Zhao G, Sun W, Yu Q (2018) Rainfall erosivity and sediment load over the Poyang Lake Basin under variable climate and human activities since the 1960s. Theor Appl Climatol:1–16

Gupta S, Kumar S (2017) Simulating climate change impact on soil erosion using RUSLE model – A case study in a watershed of mid-Himalayan landscape. J Earth Syst Sci 126:43

Gvozdetsky NA, Mikhailov NI (1978) Asian part. In: Physical geography of the USSR. 3 edn. Mysl, Moscow, p 572

Hagg W, Braun LN, Kuhn M, Nesgaard TI (2007) Modelling of hydrological response to climate change in glacierized Central Asian catchments. J Hydrol 332:40–53. https://doi.org/10.1016/j.jhydrol.2006.06.021

Hamidov A, Helming K, Balla D (2016) Impact of agricultural land use in Central Asia: a review. Agron Sustain Dev 36:6. https://doi.org/10.1007/s13593-015-0337-7

Hijmans RJ, Cameron SE, Parra JL, Jones PG, Jarvis A (2005) Very high resolution interpolated climate surfaces for global land areas. Int J Climatol 25:1965–1978

Hu Z, Zhou Q, Chen X, Qian C, Wang S, Li J (2017) Variations and changes of annual precipitation in Central Asia over the last century. Int J Climatol 37:157–170

Hu Z et al (2018) "Dry gets drier, wet gets wetter": A case study over the arid regions of central Asia. Int J Climatol

Ibatullin S, Yasinsky V, Mironenkov A (2009) Impacts of climate change on water resources in Central Asia Sector report of the Eurasian development bank 42

Immerzeel W, Pellicciotti F, Bierkens M (2013) Rising river flows throughout the twenty-first century in two Himalayan glacierized watersheds. Nat Geosci 6:742–745

Issanova G, Abuduwaili J (2017) Natural conditions of Central Asia and land-cover changes. In: Aeolian proceses as Dust Storms in the Deserts of Central Asia and Kazakhstan. Springer, pp 29-49

Issanova G, Abuduwaili J, Kaldybayev A, Semenov O, Dedova T (2015) Dust storms in Kazakhstan: frequency and division. J Geol Soc India 85:348–358

Issanova G, Jilili R, Abuduwaili J, Kaldybayev A, Saparov G, Yongxiao G (2018) Water availability and state of water resources within water-economic basins in Kazakhstan. Paddy Water Environ 16:183–191

Jahun B, Ibrahim R, Dlamini N, Musa S (2015) Review of soil erosion assessment using RUSLE model and GIS. J Biol Agric Healthcare 5:36–47

Jarvis A, Reuter HI, Nelson A, Guevara E (2008) Hole-filled SRTM for the globe Version 4

Karamage F et al (2016) Extent of cropland and related soil erosion risk in Rwanda. Sustainability 8:609

Kendall M (1975) Rank correlation methods, 4th edn. Charles Griffin, San Francisco

Kenzhebaev R, Barandun M, Kronenberg M, Chen Y, Usubaliev R, Hoelzle M (2017) Mass balance observations and reconstruction for Batysh Sook Glacier, Tien Shan, from 2004 to 2016. Cold Reg Sci Technol 135:76–89. https://doi.org/10.1016/j.coldregions.2016.12.007

Khitrov NB, Ivanov AL, Zavalin AA (2007) Problems of degradation, protection and ways of recovery productivity of agricultural land. Vestnik Orel GAU:29–32. (in Russian)

Kinnell P (2010) Event soil loss, runoff and the Universal Soil Loss Equation family of models: A review. J Hydrol 385:384–397

Kirkby MJ et al. (2004) Pan-European Soil Erosion Risk Assessment: The PESERA Map Version 1 October 2003, Explanation of Special Publication Ispra 2004 no. 73, SPI 04.73

Koshim A, Karatayev M, Clarke ML, Nock W (2018) Spatial assessment of the distribution and potential of bioenergy resources in Kazakhstan. Adv Geosci 45:217–225

Kulikov M (2018) Effects of land use and vegetation changes on soil erosion of alpine grazing lands-Fergana Range, Southern Kyrgyzstan

Kulikov M, Schickhoff U, Borchardt P (2016) Spatial and seasonal dynamics of soil loss ratio in mountain rangelands of south-western Kyrgyzstan. J Mt Sci 13:316–329. https://doi.org/10.1007/s11629-014-3393-6

Kulikov M, Schickhoff U, GrÖNgrÖFt A, Borchardt P (2020) Modelling soil erodibility in mountain rangelands of southern Kyrgyzstan. Pedosphere 30:443–456. https://doi.org/10.1016/S1002-0160(17)60402-8

Lai C, Chen X, Wang Z, Wu X, Zhao S, Wu X, Bai W (2016) Spatio-temporal variation in rainfall erosivity during 1960–2012 in the Pearl River Basin, China. CATENA 137:382–391

Lee J-H, Heo J-H (2011) Evaluation of estimation methods for rainfall erosivity based on annual precipitation in Korea. J Hydrol 409:30–48. https://doi.org/10.1016/j.jhydrol.2011.07.031

Lee E, Ahn S, Im S (2017) Estimation of soil erosion rate in the Democratic People's Republic of Korea using the RUSLE model. For Sci Technol 13:100–108

Li Z, Fang H (2016) Impacts of climate change on water erosion: A review. Earth Sci Rev 163:94–117

Liang L, Wenpeng D, Huimin Y, Lin Z, Yu D (2017) Spatio-temporal Patterns of Vegetation Change in Kazakhstan from 1982 to 2015. J Res Ecol 8:378–385

Lioubimtseva E, Henebry GM (2009) Climate and environmental change in arid Central Asia: Impacts, vulnerability, and adaptations. J Arid Environ 73:963–977. https://doi.org/10.1016/j.jaridenv.2009.04.022

Litschert SE, Theobald DM, Brown TC (2014) Effects of climate change and wildfire on soil loss in the Southern Rockies Ecoregion. Catena 118:206–219. https://doi.org/10.1016/j.catena.2014.01.007

Liu S, Duan A (2017) Impacts of the global sea surface temperature anomaly on the evolution of circulation and precipitation in East Asia on a quasi-quadrennial cycle. Clim Dyn:1–18

Luo M, Liu T, Meng F, Duan Y, Bao A, Frankl A, De Maeyer P (2019) Spatiotemporal characteristics of future changes in precipitation and temperature in Central Asia. Int J Climatol 39:1571–1588

Mamytov AM, Roychenko GI (1961) Soil zoning of Kyrgyzstan. Frunze, Izd-vo AN Kirg. (in Russian)

Mann HB (1945) Nonparametric tests against trend Econometrica. J Econom Soc:245–259

Mariotti A (2007) How ENSO impacts precipitation in southwest central Asia. Geophys Res Lett 34

McCown RL, Hammer GL, Hargreaves JNG, Holzworth DP, Freebairn DM (1996) APSIM: a novel software system for model development, model testing and simulation in agricultural systems research. Agric Syst 50:255–271

Meusburger K, Steel A, Panagos P, Montanarella L, Alewell C (2012) Spatial and temporal variability of rainfall erosivity factor for Switzerland. Hydrol Earth Syst Sci 16:167–177. https://doi.org/10.5194/hess-16-167-2012

Mitasova H, Hofierka J, Zlocha M, Iverson LR (1996) Modelling topographic potential for erosion and deposition using GIS. Int J Geogr Inf Syst 10:629–641

Mondal A, Khare D, Kundu S (2016) Change in rainfall erosivity in the past and future due to climate change in the central part of India. International Soil and Water Conservation Research 4:186–194

Morgan R et al (1998) The European Soil Erosion Model (EUROSEM): a dynamic approach for predicting sediment transport from fields and small catchments. Earth Surf Proces Landforms J Brit Geomorphol Group 23:527–544

Mukanov Y et al (2019) Estimation of annual average soil loss using the Revised Universal Soil Loss Equation (RUSLE) integrated in a Geographical Information System (GIS) of the Esil River basin (ERB). Kazakhstan Acta Geophysica. https://doi.org/10.1007/s11600-019-00288-0

Nachtergaele F et al. (2009) Harmonized world soil database (version 1.1) FAO, Rome, Italy & IIASA, Laxenburg, Austria

Nachtergaele F et al. (2010) The harmonized world soil database. In: Proceedings of the 19th World congress of soil science, soil solutions for a changing World, Brisbane, Australia, 1–6 August 2010, pp 34–37

Naipal V, Reick CH, Pongratz J, Van Oost K (2015) Improving the global applicability of the RUSLE model-adjustment of the topographical and rainfall erosivity factors. Geosci Model Dev 8:2893–2913

Nash JE, Sutcliffe JV (1970) River flow forecasting through conceptual models part I – a discussion of principles. J Hydrol 10:282–290

Nasir N, Selvakumar R (2018) Influence of land use changes on spatial erosion pattern, a time series analysis using RUSLE and GIS: the cases of Ambuliyar sub-basin. India Acta Geophysica:1–10

Nearing M (2001) Potential changes in rainfall erosivity in the US with climate change during the 21st century. J Soil Water Conserv 56:229–232

Nurbekov A, Akramkhanov A, Kassam A, Sydyk D, Ziyadaullaev Z, Lamers J (2016) Conservation Agriculture for combating land degradation in Central Asia: a synthesis. AIMS Agri Food 1:144–156

Nyesheja EM, Chen X, El-Tantawi AM, Karamage F, Mupenzi C, Nsengiyumva JB (2018) Soil erosion assessment using RUSLE model in the Congo Nile Ridge region of Rwanda. Phys Geogr:1–22

Ostovari Y, Ghorbani-Dashtaki S, Bahrami H-A, Naderi M, Dematte JAM (2017) Soil loss prediction by an integrated system using RUSLE, GIS and remote sensing in semi-arid region. Geoderma Reg 11:28–36

Panagos P, Borrelli P, Meusburger K, Alewell C, Lugato E, Montanarella L (2015a) Estimating the soil erosion cover-management factor at the European scale. Land Use Policy 48:38–50. https://doi.org/10.1016/j.landusepol.2015.05.021

Panagos P et al (2015b) The new assessment of soil loss by water erosion in Europe. Environ Sci Pol 54:438–447

Panagos P, Ballabio C, Borrelli P, Meusburger K (2016a) Spatio-temporal analysis of rainfall erosivity and erosivity density in Greece. Catena 137:161–172. https://doi.org/10.1016/j.catena.2015.09.015

Panagos P et al (2016b) Monthly rainfall erosivity: conversion factors for different time resolutions and regional assessments. Water 8:119

Panagos P, Ballabio C, Meusburger K, Spinoni J, Alewell C, Borrelli P (2017a) Towards estimates of future rainfall erosivity in Europe based on REDES and WorldClim datasets. J Hydrol 548:251–262. https://doi.org/10.1016/j.jhydrol.2017.03.006

Panagos P et al (2017b) Global rainfall erosivity assessment based on high-temporal resolution rainfall records. Sci Rep 7:4175. https://doi.org/10.1038/s41598-017-04282-8

Pimentel D (2006) Soil erosion: a food and environmental threat. Environ Dev Sustain 8:119–137

Plangoen P, Babel MS, Clemente RS, Shrestha S, Tripathi NK (2013) Simulating the impact of future land use and climate change on soil erosion and deposition in the Mae Nam Nan sub-catchment, Thailand. Sustainability 5:3244–3274

Prasannakumar V, Vijith H, Abinod S, Geetha N (2012) Estimation of soil erosion risk within a small mountainous sub-watershed in Kerala, India, using Revised Universal Soil Loss Equation (RUSLE) and geo-information technology. Geosci Front 3:209–215

Qushimov B, Ganiev I, Rustamova I, Haitov B, Islam K (2007) Land degradation by agricultural activities in Central Asia Climate Change and Terrestrial Carbon Sequestration in Central Asia. Taylor & Francis, London, pp 137–147

Ramirez-Villegas J, Jarvis A (2010) Downscaling global circulation model outputs: the delta method decision and policy analysis. Working Paper No. 1

Renard KG, Freimund JR (1994) Using monthly precipitation data to estimate the R-factor in the revised USLE. J Hydrol 157:287–306

Renard KG, Foster GR, Weesies G, McCool D, Yoder D (1997) Predicting soil erosion by water: a guide to conservation planning with the Revised Universal Soil Loss Equation (RUSLE) vol 703. United States Department of Agriculture Washington, DC

Riquetti NB, Mello CR, Beskow S, Viola MR (2020) Rainfall erosivity in South America: Current patterns and future perspectives. Sci Total Environ 724:138315. https://doi.org/10.1016/j.scitotenv.2020.138315

Sadeghi SHR, Hazbavi Z (2015) Trend analysis of the rainfall erosivity index at different time scales in Iran. Nat Hazards 77:383–404

Schmidt S, Alewell C, Meusburger K (2018) Mapping spatio-temporal dynamics of the cover and management factor (C-factor) for grasslands in Switzerland. Remote Sens Environ 211:89–104. https://doi.org/10.1016/j.rse.2018.04.008

Schmidt S, Tresch S, Meusburger K (2019) Modification of the RUSLE slope length and steepness factor (LS-factor) based on rainfall experiments at steep alpine grasslands. MethodsX 6:219–229

Shigaeva ZA (2008) Assessment of the state of land resources and land use transformation (on the example of the Sokuluk river basin). Osh Technological University

Smaling E, Fresco L (1993) A decision-support model for monitoring nutrient balances under agricultural land use (NUTMON). Geoderma 60:235–256

Sommer R, de Pauw E (2011) Organic carbon in soils of Central Asia—status quo and potentials for sequestration. Plant Soil 338:273–288. https://doi.org/10.1007/s11104-010-0479-y

Sujatha ER, Sridhar V (2018) Spatial Prediction of Erosion Risk of a Small Mountainous Watershed Using RUSLE: A Case-Study of the Palar Sub-Watershed in Kodaikanal, South India. Water 10:1608

Taylor KE, Stouffer RJ, Meehl GA (2012) An overview of CMIP5 and the experiment design. Bull Am Meteorol Soc 93:485–498

Teng H, Rossel RAV, Shi Z, Behrens T, Chappell A, Bui E (2016) Assimilating satellite imagery and visible–near infrared spectroscopy to model and map soil loss by water erosion in Australia. Environ Model Softw 77:156–167

Teng H et al (2018) Current and future assessments of soil erosion by water on the Tibetan Plateau based on RUSLE and CMIP5 climate models. Sci Total Environ 635:673–686

Teng H-f, Hu J, Zhou Y, Zhou L-q, Shi Z (2019) Modelling and mapping soil erosion potential in China. J Integr Agric 18:251–264. https://doi.org/10.1016/S2095-3119(18)62045-3

Thomas J, Joseph S, Thrivikramji K (2018a) Assessment of soil erosion in a tropical mountain river basin of the southern Western Ghats, India using RUSLE and GIS. Geosci Front

Thomas J, Joseph S, Thrivikramji K (2018b) Estimation of soil erosion in a rain shadow river basin in the southern Western Ghats, India using RUSLE and transport limited sediment delivery function. International Soil and Water Conservation Research

Tucker CJ (1978) Red and photographic infrared linear combinations for monitoring vegetation

Uddin K, Murthy M, Wahid SM, Matin MA (2016) Estimation of soil erosion dynamics in the Koshi basin using GIS and remote sensing to assess priority areas for conservation. PloS ONE 11:e0150494

Unger-Shayesteh K, Vorogushyn S, Merz B, Frede H-G (2013) Introduction to "water in Central Asia—Perspectives under global change". Glob Planet Chang:1–3

Van Diepen C, Wolf J, Van Keulen H, Rappoldt C (1989) WOFOST: a simulation model of crop production. Soil Use Manag 5:16–24

Wall G, Coote D, Pringle E, Shelton I (2002) Revised Universal Soil Loss Equation for Application in Canada: A Handbook for Estimating Soil Loss from Water Erosion in Canada. Agriculture and Agri-Food Canada, Research Branch, Ottawa. Contribution No AAFC/AAC2244E

Williams JR (1975) Sediment-yield prediction with universal equation using runoff energy factor

Williams M, Konovalov V (2008) Central Asia temperature and precipitation data, 1879–2003. USA National Snow and Ice Data Center, Boulder

Wischmeier W, Smith D (1965) Rainfall-erosion losses from cropland east of the Rocky Mountains, guide for selection of practices for soil and water conservation Agriculture Handbook 282

Wischmeier WH, Smith DD (1978) Predicting rainfall erosion losses-a guide to conservation planning Predicting rainfall erosion losses-a guide to conservation planning

Yang D, Kanae S, Oki T, Koike T, Musiake K (2003) Global potential soil erosion with reference to land use and climate changes. Hydrol Process 17:2913–2928

Zafirah N, Nurin N, Samsurijan M, Zuknik M, Rafatullah M, Syakir M (2017) Sustainable ecosystem services framework for tropical catchment management: a review. Sustainability 9:546

Zhang G et al (2018) Exacerbated grassland degradation and desertification in Central Asia during 2000–2014. Ecol Appl 28:442–456

Index

E. Duulatov et al., *Current and Future Trends of Rainfall Erosivity and Soil
Erosion in Central Asia*, SpringerBriefs in Environmental Science,
https://doi.org/10.1007/978-3-030-63509-1

Printed in the United States
By Bookmasters